The Hidden Mechanics
of Exercise

The Hidden Mechanics of Exercise

Molecules That Move Us

Christopher M. Gillen

THE BELKNAP PRESS OF HARVARD UNIVERSITY PRESS

Cambridge, Massachusetts
London, England
2014

Library of Congress Cataloging-in-Publication Data

Gillen, Christopher M., author.
    The hidden mechanics of exercise : molecules that move us /
Christopher M. Gillen.
        pages   cm
Includes bibliographical references and index.
ISBN 978-0-674-72494-5 (alk. paper)
    1. Exercise—Physiological aspects.   2. Exercise—Molecular aspects.
3. Human mechanics.   4. Kinesiology.   I. Title.
QP303.G543   2014
612.7'6—dc23   2013034257

I dedicate this book to my mother and father,
who inspired me to exercise my mind, body, and spirit.

Contents

The Hidden Mechanics
of Exercise

Molecules in Motion

At a local park on a warm summer evening, a teenager nails a three-pointer while her parents complete a long run. A skateboarder grinds across a rail while his grandmother crushes a backhand down the line. A shortstop nabs a line drive while her brother drills a penalty kick into the lower corner. These routine athletic movements demand power, coordination, and control. If we carefully watch athletes—even amateurs at the gym or on the playing fields—we see synchronized limbs, balanced bodies, and precise timing. We also observe occasional moments of clumsiness and frustration, highlighting the difficulty of everyday athletic moves.

But no matter how diligently we observe, the molecules that motivate these athletic actions remain hidden. We don't see ions whizzing across membranes, motor proteins tugging on molecular filaments, or enzymes extracting energy from food. We don't see receptors morph, carrying signals into cells. We don't see proteins pirouette, converting glucose into muscle power. We don't see the sifting of sensory information, the decision-making process, the regulation of contractions, or the allocation of energy to meet demand.

Today's physiologists, on the other hand, are fast unveiling the concealed inner workings of the athletic body. We now study exercise not just at the whole body or organ system levels, but also at the cellular and molecular levels. We know that harmonized movements of myriad tiny molecules parallel the visible grace of athletes. The connection between this molecular ballet and athletic performance is the subject of this book.

Modern biology textbooks portray molecules in striking detail. Their illustrations are not vague cartoons of what the molecules might look like, but accurate portrayals of the miniscule structures that compose our bodies. Unlocking the form of biomolecules requires extraordinary technical prowess, so textbook images of molecules depict remarkable scientific achievements. Yet the flat, static renderings in textbooks do little justice to the grandeur of biological molecules in the same way that photographs fail to capture the magnificence of a soaring, two-handed, reverse slam dunk. Biomolecules are not passive entities bumping around in our bodies. Rather, they are active, kinetic, dynamic shape-shifters, morphing into different conformations in the course of their duties. The well-timed muscular contractions of a slam dunk, for example, rely upon the synchronized contortions of countless molecules. In other words, molecules in motion power bodies in motion.

Consider hemoglobin, one of the best-studied proteins. On the page, hemoglobin's structure—four blobs, each cradling an iron atom—is rather uninspiring. But set hemoglobin awhirl in human blood, and its splendor surfaces. Hemoglobin pulsates, relaxing and tensing as it binds and releases oxygen. And it jiggles, subtly morphing in response to acidity, carbon dioxide concentration, and other factors. Yet we will miss hemoglobin's full glory if we study it in an inactive human, just as the constraints of city streets mask a sports car's potential. In Chapter 8, we will see that hemoglobin's most marvelous behavior occurs during exercise.

Two strands of my life meet in this book: my career as a scientist and my passion for distance running. My scientific background includes training in both exercise physiology and molecular biology. My running résumé lists over two dozen marathons and a handful of ultramarathons. A recent race experience illustrates how the connections between science and exercise have enriched my own athletic pursuits and explains what I hope readers will gain from this book.

The fourth time I attempted the Mohican 100-mile trail run, I stood confidently on the starting line. My prior outings had been unimpressive. Twice I dropped out well before the finish; once I straggled across the finish line dragging an injured leg. But in the year preceding my fourth attempt, I systematically addressed the failings of previous years. Sleepiness had been a problem, especially in the nighttime portion of the race, so I planned to start drinking caffeinated beverages in the late afternoon. Soreness of my iliotibial band (ITB)—a thin muscle that runs from hip to knee—had often plagued me, so in the months leading up to the race I stretched and strengthened that muscle. Just in case, I wore a tight strap above my knee to relieve strain on the ITB. And finally, dehydration and heat stress had slowed me in past tries, so I carried a water bottle in each hand and two types of salt tablets in my pockets. (Chapter 10, which discusses sports beverages, describes my rationale for taking salt during the race.)

My solutions to each of these problems came partly from the advice of other runners and partly from my own experiments on the trails, but also from my knowledge of exercise science. Athletes can certainly learn from their own experiences and those of others, but these experiences must be applied cautiously to new situations. We all have different bodies, so what worked for some may not work for others. And conditions differ from one event to another, so what worked yesterday may not work today. Knowing how our bodies function can help us apply previous experiences to the current situation.

This book's goal—connecting molecules to motion—offers practical rewards because the body's molecules hold answers to questions that athletes routinely ask. Should I take salt tablets? Why does exercise make me feel good? Do I need another recovery day? What should I eat before the race? The answers hinge on our molecular makeup. Of course, many athletes get along just fine without knowing about biomolecules in the same way that many drivers get along without knowing how a transmission works. On the other hand, you can be sure that NASCAR racers know the difference between a carburetor and a crankshaft. Likewise, if you want to maximize your athletic success, it helps to know something about molecular mechanisms.

Let's consider the protein myosin, the workhorse of muscle cells. Myosin has a thin tail separated from a globular head by a small flexible neck. During muscle contraction, myosin heads bind energy-rich adenosine triphosphate (ATP) molecules, break them apart, and capture the released energy. This energy moves the heads, activating them like stretched springs. The energized heads grab other muscle proteins and snap back to their original position, generating muscular power. So myosin—like hemoglobin—

morphs. Myosin's contortionist skills, as we will see in Chapter 4, hold the key to practical questions such as "How heavy should a baseball bat be?"

Though athletes will find this book useful, it is not a training guide, lifestyle manual, or diet manifesto. The point is not to tell you what to do, but to equip you to seek your own answers. In the language of a legal disclaimer: no part of this book should be construed as health or training advice. On the other hand, the book will help you interpret reports about health and fitness in the popular media, ask good questions of health care professionals and coaches, and make your own well-informed decisions.

I have a second, less utilitarian, aim for this book: I want readers to marvel with me at the magnificence of our molecules in motion. In other words, I seek to convey the stunning beauty and astonishing complexity of our molecular systems, not simply their practical applications. This approach mirrors my rationale for running. I certainly run for the practical benefits—fitness, weight control, stress relief. But I also appreciate the intangible joys of running, as the following continuation of the story about my fourth Mohican 100 illustrates.

More than twenty hours after leaving the starting line, long before the first trace of dawn, I walked briskly with a friend, our headlamps illuminating only tiny fragments of the dense surrounding forest. In a clearing, we switched off the lights, felt the darkness embrace us, then looked up and watched a shooting star flash across the sky. Were we hallucinating? I don't think so because the volunteers staffing the next aid station also claimed to have seen the shooting star. On the other hand, maybe we were seeing things and the volunteers were just humoring us—exercise triggers the same neural pathways as certain recreational drugs, according to recent research that we discuss in Chapter 7. Whatever their cause, moments like these partly explain why I run ultramarathons. I treasure the way running unites me with my surroundings and with people, both other runners and race volunteers. These connections, I realize, are not unique to running—skiers, weightlifters, softball players, and other athletes feel them.

I also cherish how running connects me to my own body. Shortly after seeing the shooting star, I decided to jog a short section of flat trail. After miles of walking, running felt wonderfully refreshing. The night was quiet, my headlamp lit only the ground around my feet, and my focus turned progressively inward, not only to my lungs, heart, calves, and thighs, but also to myosin, hemoglobin, and the other molecules driving my body forward. My body felt like a powerful machine, and I reveled in the majesty of the molecular movements coordinating the rhythm of my stride.

Then a stick grabbed my foot, interrupting my trance and somersaulting me into the high grass alongside the trail. I landed hard on my shoulder but bounced up largely unscathed, shaken yet not injured. I was lucky. Had I fallen differently, my race might have abruptly ended, a year of training evaporated by a moment of lost focus. I tell this part of the story—rather than leaving you with the vision of blissful runners gliding through the nighttime forest—for two reasons.

First, my tumble reflects the modest course of my athletic career. I am an eager but average athlete who has never stood on a podium, cashed a sponsor's check, or led teammates to a championship against overwhelming odds. I certainly have not scaled Everest or crossed the Sahara. Thus I occasionally tell personal anecdotes not to tout my athletic prowess but to show how molecular science connects to real-world athletics and to encourage readers to look for connections to their own experiences. Although I'm now a runner, I've dabbled in many sports—golf, tennis, skiing, soccer, ice hockey—and I'll pull examples from these and other sports too. As a rule, my focus will remain on everyday athletes, not professionals. This book is not about superstars; it's about the rest of us.

Second, in this book I plan to avoid making the error that resulted in my hard fall. For a moment, I lost contact with the outside world, completely neglecting practical issues such as where my foot was going to land. Scientists sometimes revel in an abstract, theoretical world—indeed, we must sometimes inhabit this world—in much the same way that athletes sometimes focus on their own body mechanics. In this book, I will occasionally stop to gaze at shooting stars, to wonder at the grace of human bodies in motion, to admire the beauty and precision of the molecules that move us. But I promise not to get lost in the abstract molecular world. Practical real-world issues that interest athletes permeate every chapter.

In fact, though I've offered two very different rationales for reading this book—one tangible and practical, the other intangible and aesthetic—these rationales are intertwined. Just as an alley-oop dunk inspires awe because multiple events—pass, leap, catch, slam—are perfectly timed to score a bucket, the activities of myosin—catalyze, energize, bind, morph—are amazing because together they achieve muscular contraction. Our molecular machinery is superb, especially when viewed through the lens of its function. Of course, my brief description in this Prologue fails to capture the true glory of myosin, just as my explanation of the joys of ultrarunning falls short. To appreciate the thrill of trail running, you need to run the trails yourself or, preferably, with an experienced companion. And to grasp the full grandeur of myosin and other biomolecules, you need to learn more about them, either on your own or, better yet, by joining me on this book's journey.

While I was writing this book, some told me that molecular intricacies were too complex for the general reader, that I could either dumb down the material or narrow my audience to a few scientifically trained athletes. I disagree. But I do admit that molecular biology can be difficult—something I learned the hard way when I joined a molecular biology laboratory after doing graduate work in exercise physiology. Initially, I felt a bit like a beginner on a tennis court—inept, disoriented, and frustrated. Like tennis, molecular biology takes considerable patience and practice to master. Just as a good coach can accelerate your progress in tennis, however, I can help newcomers to biomolecules quickly become familiar with the molecular world by employing several tactics.

First, I encourage you to link biology to your own athletic endeavors. When you begin to connect the molecular science to concrete experiences, apprehensiveness often transforms into exhilaration. In other words, looking at the practical side of molecules offers a window into their deep intrinsic splendor. This complements the point I made previously: knowing about biomolecules can enrich athletics. Now I'm making the reciprocal argument: a good way to understand biomolecules is to explore their roles in exercise.

Scientists studying molecules sometimes seem to speak a different language—appropriate to their discipline but off-putting to others. I shy away from this technical jargon whenever possible. Yet discussions about terminology occupy us throughout this book. In many cases, a term encompasses a concept that requires a sentence or two to explain in plain language. To write efficiently, I sometimes need to use such terms, but I always carefully define them in the text and Glossary. We will also see that scientific controversies—such as those about lactic acid described in Chapter 5—sometimes pivot on arguments about the meaning of terms, another reason to pay attention to terminology.

To a certain degree, we should expect some bewilderment when we contemplate a world beyond our immediate senses. We have difficulty envisioning how molecules work because they are too small to observe with our unaided eyes. So throughout this book, I employ analogies to help you "see" the molecular world. I compare proteins to kitchen utensils, enzymes to referees, ions to ants, and cell layers to castle walls. Though such analogies help us comprehend a world beyond our immediate senses, they inevitably fail to perfectly capture biological reality. Consider my analogies as tools, but don't take them too literally. Computer visualizations are also valuable tools that allow users to examine structures, rotate molecules, zoom in on regions of interest, and highlight different features. The Appendix describes how readers can find and use free software to explore the biomolecules described in this book. I encourage you to take advantage of these computer programs.

Each of the following chapters explores the molecular basis of key aspects of exercise physiology. Chapter 1 investigates how a protein called collagen enables us to move smoothly and efficiently. Collagen's properties will help us ponder whether to stretch and whether to run barefoot. In Chapter 2, we examine the enzyme creatine kinase. This enzyme's role in providing energy to exercising muscles sheds light on the value of the performance-enhancing supplement creatine. The possibility of using alpha-actinin-3 (ACTN3) to predict athletic success is the topic of Chapter 3. We see in Chapter 4 how the structure and function of myosin, the protein at the heart of muscle contraction, explains trade-offs in muscle properties. For example, the interaction of myosin with other muscle proteins clarifies why we cannot simultaneously maximize both muscle force and shortening velocity. Lactic acid takes center stage in Chapter 5, where I describe recent research showing that lactic acid plays a helpful role during exercise rather than causing suffering and fatigue.

Chapters 6 and 7 investigate how the brain and nerves control muscle contractions. In Chapter 6, we explore how proteins that allow salt to flow into and out of cells contribute to the reflex pathways that coordinate movements such as snowboard jumps. In Chapter 7, we see how exercise influences the brain by modulating communications between nerve cells. We explore immediate effects such as runner's high as well as long-term effects of exercise on memory and other processes. Aerobic metabolism is the topic of Chapter 8. We discuss how environmental factors such as altitude exposure influence the levels of key proteins such as hemoglobin. Chapter 9 describes physiological differences between men and women, focusing on the ratio of fat and carbohydrates as energy sources during exercise. We explore the enzymes and hormones that may contribute to these sex differences.

The remaining chapters show how multiple molecules work together to achieve complex functions during exercise. In Chapter 10, we look at the value of sports drinks in maintaining fluid balance during exercise. This chapter highlights how proteins collaborate in cells and tissues of the kidney, sweat glands, and intestine. Ribonucleic acid (RNA) is the topic of Chapter 11. We explore the role of different types of RNA in exercise training, and see that RNA not only carries information to make proteins but also catalyzes and regulates reactions. Finally, Chapter 12 examines new research into the genetic contributions to exercise. We discuss new techniques that allow scientists to simultaneously evaluate every gene in our genome and ponder the significance of these developments for athletes.

Each chapter employs examples drawn from specific sports, but the mechanisms we will discuss operate in every athlete's body. Weightlifters and water polo players use motor proteins in the same way to produce force. Badminton players and bicyclists send signals along nerves with the same type of salt channels. And triathletes and trail runners extract energy from food by the same metabolic pathways. So whatever your sport, an understanding of molecular mechanisms will help you squeeze a bit more joy from your athletics and perhaps even improve your performance the next time you compete.

Chapter 1

Function Follows Form

Tension and tightness dissolve a mile into the run. My breathing evens out. My heart beats an easy cadence. My calm, detached mind enjoys the ride without interfering. I float along the bike path, legs limber, feet brushing the ground lightly, each step propelling the next, my movements effortless, integrated, exhilarating. I am—at least for the moment and at least in my own mind—a fluid, graceful machine.

Efficient human movements require resilience, the physiological ability to store and return energy. Without resilient tendons, the threads of tissue that connect muscles to bones, human movement would be jerky and wasteful. Tendons specialize in resilience, but they also must be strong and stiff—able to handle heavy loads and resist changes in length. Strength, stiffness, and resilience arise from the unique structure of collagen, the main component of tendons. Other tissues require different properties. Arteries, the vessels that carry blood away from the heart, share with tendons the property of resilience, but must also be extensible— able to stretch without snapping. Their extensibility comes from a molecule called elastin. Just as the physical structures of muscles and bones influence their function, the chemical structures of collagen and elastin mold their functional properties.

When we exercise, we move bones, changing their relationship to each other. The elbow joint begins a biceps curl in an extended position, with bones of the forearm and upper arm aligned. As the biceps muscle shortens, the elbow joint flexes, bringing the forearm toward the shoulder. More complicated athletic actions—such as triple axels, baseball swings, and pole vaults—may require hundreds of separate bone movements. But all these maneuvers ultimately reduce to a series of movements of one bone with respect to another.

Muscles and tendons partner to move bones during exercise—muscles generate force, tendons transmit the force to bones. Although muscles may attract more attention at the beach and more ink in the fitness magazines, tendons are equally essential to successful movements. Intermediaries between muscles and bones, tendons link muscular force to physical action. Tendons attach each muscle to at least two different bones. For example, tendons join one end of the biceps muscle to the shoulder blade and the other end to the forearm near the elbow. Because of these connections, the forearm moves toward the shoulder when the biceps shortens.

Throughout history, humans have exploited the remarkable properties of tendons. The armies of Dionysius the Elder likely used crossbows reinforced by animal tendons as early as 399 BC. Writing in the first century AD, Hero of Alexandria described the use of tendons in catapult torsion springs. And for centuries, people have fashioned animal tendons into bowstrings. An effective bowstring resists breaking when pulled back, transmits force without large changes in its length, and stores energy when stretched. In other words, bowstrings are strong, stiff, and resilient, so the qualities that make tendons effective transmitters of force from muscle to bone also make tendons good bowstrings. As I said earlier, the properties of tendons come from the molecular makeup of collagen. Collagen, which accounts for 80 percent of the dry weight of tendons, belongs to a class of molecules that will engage our attention throughout this book: proteins.[1]

Proteins are tiny tools, devices, and machines, analogous to the gadgets that pack a typical kitchen. Just as kitchen tools—pans, knives, blenders, bread machines—perform various tasks, proteins play varied roles in our bodies. Some proteins regulate cell processes; others form cell structures. Some proteins produce mechanical force; others catalyze chemical reactions. Some proteins move molecules in and out of cells; others drag molecules from one place to another inside cells. Almost all proteins interact with other molecules—including other proteins—in the course of their duties, just as a knife and cutting board collaborate to dice vegetables. And just as a blender can be used for countless purposes, many proteins perform multiple functions.

Form determines the effectiveness of both kitchen implements and proteins. We grasp knives, forks, and spoons by shafts that have the same shape because they serve the same role—they are handles. Other parts of each utensil reflect a specific function: knife blades cut, fork tines spear, and spoon bowls scoop. Within each category of utensil, structures vary to match tasks. Butter spreaders, steak knives, and vegetable cleavers modify the basic knife design to accomplish specialized functions. As with utensils, the three-dimensional structure of proteins dictates their capabilities. Some proteins, the teaspoons of the protein world, have simple structures. Other proteins are more like bread machines, complex assemblages with myriad moving parts. Different proteins share common structural motifs with widely useful functions in the same way that different utensils share the same handle design. At the same time, many proteins contain unique structural regions necessary for distinct functions, much like blades, bowls, and tines of utensils. In short, proteins flaunt an extravagant array of shapes. Collagen assembles into ropelike strands. Other proteins are oval, helical, planar, globular, filamentous, and just about every other shape you can imagine.

Manufacturers fashion kitchen utensils using a variety of materials and processes. In contrast, all proteins share the same building blocks. To make proteins, cells link simple molecules called amino acids into chains that fold into innumerable shapes. Envision a protein's amino acids as beads on a necklace. Then imagine a necklace comprising twenty different kinds of beads, equivalent to the number of different amino acids. We can fashion 8,000 different necklaces or proteins by linking just three beads or amino acids ($20 \times 20 \times 20 = 20^3 = 8,000$). Necklaces of 200 beads—the number of amino acids in a typical protein—can come in 20^{200} or 10^{260} (10 followed by 259 zeros) combinations. Thus a stunning variety of necklaces—and proteins—is possible. Now, imagine that some beads are magnetic, so they attract or repel other beads, twisting the necklace into a complex shape. Meanwhile, large beads constrain the necklace, preventing it from bending certain ways. The sequence of beads determines exactly how the necklace folds; if two nearby beads attract each other, the necklace will twist differently than if two distant beads attract. The shape of our imaginary necklace—how it coils and contorts—depends on the properties and sequence of its beads.

Likewise, protein structure arises from the properties and sequence of amino acid chains. Amino acids consist of a central carbon atom connected to four groups (Figure 1.1). Three of these groups—the hydrogen atom, the carboxylic acid group, and the amino group—are invariant across all amino acids. The fourth group—the side group—varies in shape, electrical charge, chemical reactivity, and solubility in water. A protein's linear sequence of amino acids determines how it folds. For example, side groups with positive charges repulse other positively charged groups but attract groups with negative charges. Similarly, water-soluble amino acids often line protein surfaces, facing outward toward the surrounding water. Although factors such as temperature and acidity also influence protein shape, amino acid sequence plays the leading role. The nearly limitless permutations of amino acid sequence make possible the dazzling diversity of protein forms.[2]

1.1. Amino acid chemistry. Above: Chemical structure of an amino acid. Below: Four amino acids joined into a peptide. Arrows point to peptide bonds.

Collagen's amino acid sequence displays a distinctive pattern: many repeats of a three-amino-acid motif, represented as Gly-X-Y. Each triplet starts with the small amino acid glycine (Gly). Though the remaining positions (X and Y) can be any amino acid, the bulky amino acid proline (or the related molecule hydroxyproline) often fills them. How does collagen's repeating motif relate to the unique properties of tendons? We will answer this question first with respect to strength and stiffness, and then turn to resilience.

Many modern athletes can sympathize with Achilles, the mythical Greek hero. Achilles's mother held his heel when she dipped him in the River Styx, making his entire body except that heel invincible. Achilles eventually died, as the story goes, when a poisoned arrow lodged in his susceptible heel. Though today's athletes rarely need to dodge poison arrows, vulnerable heels still plague us. Many sports require movements that seriously stress the aptly named Achilles tendon, which connects calf muscles to the calcaneus bone of the heel.[3]

Forces on the Achilles peak during sprinting, reaching almost ten times body weight, about 1,500 pounds in an adult. Yet, though it is thicker than any other tendon in our body, the Achilles is surprisingly thin, with a diameter between 6 and 7 millimeters (mm), about the width of a standard pencil. Consequently, the cross-sectional area of the Achilles is about a hundred-fold less than that of the calf muscle it serves. The Achilles's slim profile affords it access to the compact ankle joint. A skinny Achilles also lightens the foot, saving energy because weight at the end of a limb is especially burdensome to move. High forces acting on a thin tendon means that the Achilles must be particularly strong.

Imagine a Toyota Prius, which weighs about 3,000 pounds, suspended from two pencil-thin Achilles tendons, and you get a sense of the Achilles tendon's strength. Though they occasionally fail, Achilles tendons can withstand high forces before they break.[4]

The Achilles must not only be strong, but also stiff—it must not stretch too much when force is applied. To see why, test the mobility of your ankle. Extend the ankle by contracting your calf, pushing your toes downward toward the sole of your foot and away from the shinbone (Figure 1.2). Then, flex the ankle by pulling your toes upwards toward the shinbone, and feel the Achilles tendon stretch.[5] Your foot follows a similar pattern during running, hitting the ground in a relaxed, somewhat extended

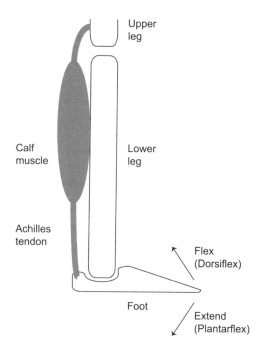

1.2. Diagram of lower leg and foot.

position, then flexing as weight rolls onto it. The ankle's modest range of motion matches the properties of the Achilles. About 250 mm long, the Achilles tendon is a bit longer than a standard pencil. When the ankle flexes, the Achilles extends by about 6 percent. Thus, despite bearing large forces during running, the Achilles tendon stretches only a small amount. Stiff tendons effectively transmit muscular force from the calf to the heel. When tendons lose their stiffness, muscular force stretches the tendon rather than moving the bone, impairing force transfer.

Let's return to collagen's structure to see where tendons get their strength and stiffness. Chemical bonds hold together the individual amino acids in collagen and other proteins. Two amino acids attach when the carboxylic acid group of one amino acid reacts with the amino group of another, forming a covalent bond (Figure 1.1). The atoms that participate in covalent bonds share electrons, gluing the atoms together strongly. Hence, protein chains are tough to break. Still, when exposed to forces like those on the Achilles, a single collagen strand would snap like a line of spider silk stretched across the Indy 500 finish line. To achieve adequate strength, many collagen molecules must cooperate.

Each collagen molecule contains about a thousand amino acids with many Gly-X-Y repeats. The tiny side groups of glycine, each a single hydrogen atom, form notches between the larger side groups in the X and Y positions. Three collagen threads click together at these indentations, forming a tightly wound triple helix.[6] Amino acids at the X and Y positions further stabilize the helix by forging hydrogen bonds between separate strands. Hydrogen bonds form due to the attraction between partial negative and positive changes. Although weak compared to covalent bonds, many hydrogen bonds combine to brace the helix. Weaving three strands together enhances strength. Notches and hydrogen bonds prevent strands from sliding alongside each other, contributing to the stiffness of the triple helix and making it tougher to stretch.[7]

The collagen triple helix resembles rope. Both draw strength from the braiding of individual strands. Both require strands to be held together to prevent slippage when a force is applied. And both are strong in tension when their ends are pulled apart, but not as effective in opposing compression when their ends are pushed together.

Though stronger than individual collagen molecules, triple helices are still much too weak to transduce muscular force. Each helix is only 1.3 nanometers (nm) thick and 300 nm long. Nanometers are definitely small—one nanometer is 0.00000004 inches—but it can be hard to picture lengths that short, so let's pause for a moment and try to envision what these measurements mean. Nanometers are a thousand-fold smaller than micrometers (μm), which in turn are a thousand-fold smaller than millimeters (mm). Thus nanometers are six orders of magnitude smaller than the smallest intervals on a metric ruler. To think of it another way, millimeters eclipse nanometers to the same extent that kilometers eclipse millimeters. Any way we slice it, many tiny collagen triple helices must be packed together to form a 6 mm-by-250 mm Achilles tendon.

Cells build tendons by secreting collagen triple helices into the extracellular matrix, a mixture of proteins and sugars outside cells. At this site, hundreds of triple helices group into fibrils, held tightly together by covalent bonds between amino acids at the X or Y position of Gly-X-Y triplets. In turn, fibrils aggregate into tendons with thousands of robustly attached collagen molecules running in parallel. In summary, the strength of tendons comes not from the individual talents of collagen molecules, but rather from their knack for collaboration.

Despite collagen's strength, painful injuries to the Achilles tendon commonly afflict recreational athletes. The worst type of Achilles damage, a partial or complete rupture of the tendon, sometimes triggers a sharp popping sensation that victims compare to a gunshot. Nobody wants to be hobbled by an Achilles injury, so runners learn to respect and nurture their Achilles tendons. Before and after workouts, you'll find us pushing against a wall, one leg extended back. We first straighten the extended leg to stretch our calves and then slowly bend at the knee to gently tug on our Achilles tendons. The idea is to stretch the muscles and tendons of the lower leg, decreasing their stiffness, relieving some of the stress on the Achilles.

Whether our rituals improve performance or reduce injury is a matter of some debate. Stretching does decrease the stiffness and increase the length of muscle-tendon units, which could decrease injury risk. But most controlled studies of stretching fail to demonstrate benefits. Even worse, muscle strength often decreases directly after static stretching—holding positions for fifteen or more seconds—impairing performance in activities that demand strength. One reason for these strength declines is that reduced stiffness delays transmission of force from muscles to bone. Other factors, such as impaired activation of muscles by nerves, also contribute. Regardless of the mechanisms, static stretching prior to exercise, especially exercise that requires powerful contractions, may be counterproductive. Some research suggests that dynamic stretching, which uses sports-specific movements rather than held poses, may be more beneficial than static stretching.[8]

In contrast to its negative effects on strength, stretching improves range of motion and flexibility, thereby enhancing activities such as dancing where those qualities matter. Lowering stiffness also tends to improve resilience, which could aid activities where energy is stored and returned, such as running. Conflict about the value of stretching flourishes in the scientific literature, in part because so many variables complicate the analysis. Researchers need to tease apart effects of factors including the type and timing of stretching, age and fitness of athletes, and duration and intensity of exercise. In the meantime, I will continue pushing against walls to stretch my calves. But I'm increasingly inclined to avoid stretching right before workouts and tempted to experiment with some dynamic stretches.

This section has covered a lot of ground, connecting submicroscopic amino acids to tendons that you can reach out and feel. The following are two take-home messages: First, many biological structures, including tendons, are arranged hierarchically. Amino acids assemble into collagen; collagen links into triple helices; triple helices join to form fibrils, and fibrils collaborate to make tendons. Complexity mounts at each level, as subunits join into increasingly sophisticated assemblies. Second, amino acid sequence determines each protein's form and therefore its function. Glycines form the notches that clasp triple helices together, whereas proline and other amino acids stabilize helices and fibrils by establishing chemical bonds between collagen strands. Replacing these amino acids with ones having different properties could disrupt collagen's shape and thereby degrade its function.

Although tendons display impressive strength and stiffness, these properties do not set them apart from other natural or man-made materials. For example, some spider silks are ten-fold stronger and stiffer than tendons on a per-weight basis. Tendons are also less stiff than bone, which is able to withstand high forces with very little deformation. And high-tensile steel is ten-fold stronger and a hundred-fold stiffer than tendons. But no material stores elastic potential energy more effectively than tendons. Although spring steel has a higher resilience, it is also much heavier. As a result, an ounce of tendon stores and returns energy over ten-fold more effectively than an ounce of spring steel.[9]

Highly resilient tendons smooth human movements, especially running, because they mediate transitions between kinetic energy (the energy of motion) and potential energy (stored energy). For a simple demonstration of energy transitions, toss a ball into the air. As the ball rises, kinetic energy gradually converts to gravitational potential energy. At the top of its arc, the ball reaches its farthest point from the ground and stops moving upward, achieving its maximum gravitational energy and lowest kinetic energy. At this point, the ball starts falling and converts its gravitational energy back to kinetic energy, this time in the downward direction. Runners go through the same transitions. Though we run to achieve forward movement, most of the muscular force we exert yields upward movements that temporarily elevate us off the earth so that we avoid dragging along the pavement. Upward kinetic energy transfers to gravitational energy, and we soar above the earth with both feet in the air. Then the gravitational energy converts back to kinetic energy and pulls us swiftly back to earth.

What happens when a ball hits the ground? That depends on the ball. A rubber ball deforms, stores energy, and then rebounds, springing back upwards. Kinetic energy converts to elastic potential energy, then back to kinetic energy. The rubber ball stores and returns energy because it is highly resilient. Super-bouncy balls have resiliencies in the neighborhood of 90 percent. When they hit the ground and reverse direction, they rebound with 90 percent of the original kinetic energy, losing only 10 percent in the transition. In contrast, snowballs splat upon contact because their resilience is close to zero. The kinetic energy of a snowball dissipates as heat when it collides with the ground, so getting the snowball airborne again demands another source of energy. Clearly, we want our legs to act like superballs, launching us back into the air, rather than like snowballs, pasting us to the ground.

Tendons surpass superballs in resilience. The Achilles tendon, with a resilience of about 93 percent, contributes to smooth running. When the ankle flexes during foot strike, the foot decelerates, losing both forward and downward velocity. So the Achilles stretches, converting kinetic energy into elastic potential energy. This elastic energy converts back to kinetic energy as the Achilles shortens during push-off, adding power to the push. Collagen-rich ligaments—tissue strands that attach bones to each other—in the foot's arch work in a similar way. The arch compresses during foot strike, stretching ligaments that run along the bottom of the foot, and then springs back on push-off, returning about 80 percent of the energy. When we run, structures with lower resilience dissipate some energy, so we don't achieve the 80–95 percent energy return promised by tendons and ligaments alone. But our legs still store and return about 50 percent of the total energy, so we need that much less energy to produce muscular force during the push-off. The saved energy allows us to run longer or perhaps faster.[10]

The amino acid sequence of collagen underpins the resilience of tendons. According to studies that model collagen's response to stretch, energy dissipates when adjacent collagen triple helices slip past each other, emphasizing the importance of strong connections between parallel collagens. In contrast, when links between triple helices hold firm, the helices themselves stretch, storing energy. Regions of collagen rich in proline and its derivatives are rigid because the bulky structure of these amino acids constrains movement. But regions rich in other amino acids flex like springs. Kinked at rest, these flexible regions store energy when straightened by tension, and return energy when they spring back to the kinked position.[11]

Human Achilles tendon anatomy hints that we are natural runners. We possess particularly well-developed Achilles tendons compared to chimpanzees, our nearest primate relatives, whose Achilles tendons are relatively shorter—and therefore less resilient—than ours. Consequently, chimpanzees might sprint a few yards, but you won't find them running marathons. The Achilles tendons of the extinct hominid *Australopithecus* probably resembled chimpanzee tendons more than human tendons, though interpreting fossil evidence of muscle and tendon anatomy can be tricky. Thus the elongated Achilles of humans probably arose late in our evolutionary history, perhaps as a specialization for distance running. Other evidence from diverse disciplines—anthropology, archeology, physiology, and mechanics—supports the notion that our human ancestors routinely ran long distances. Although some nonrunners may view distance running as a crazy masochistic aberration, this recent research suggests it may instead be a fulfillment of a deeply distinctive human capacity.[12]

Is it possible that our Achilles anatomy is a general adaptation for bipedalism rather than a specific one for running? Probably not. Across a range of animal groups, elongated tendons in the lower leg correlate with efficient running. For example, horses have a long skinny tendon in their lower legs, though their limb anatomy is otherwise quite different from ours. Also, we store less elastic energy during walking than in running, so our Achilles is probably not specialized specifically for walking. Finally, the properties of the human Achilles promote energy efficiency more than strength, speed, or power, suggesting that the Achilles is specifically an adaptation for running rather than explosive one-time motions like jumping. Indeed, thin lengthy tendons like our Achilles maximize energy return—improving efficiency and endurance—but are more susceptible to rupture during maximal efforts.[13]

If we are born to run, then why do we need running shoes? Why put a layer of spongy technology between our feet and the ground? A growing barefoot-running culture challenges the value of running shoes. Christopher McDougall's best-selling *Born to Run* fueled the barefoot movement by telling the tale of the Tara-humara Indians of northern Mexico, who run ultralong distances wearing only sandals.[14] According to barefoot-running enthusiasts, modern running shoes impair our natural running motion. By providing artificial support, shoes may encourage atrophy of muscles that normally stabilize our feet and legs, thereby increasing injury risk. In addition, shod runners tend to dissipate energy by landing on their heels, whereas barefoot runners typically avoid the jolt of the heel strike, landing on their fore- or mid-foot, storing energy in their resilient Achilles tendon and arch. Running shoes do not return as much energy as tendons, so barefoot running could improve running efficiency. Though these technical arguments for barefoot running inspire debate among runners and scientists, the pleasure of barefoot running is inarguable. Try running barefoot through soft grass. The abundant sensory receptors in your feet will hum joyfully and you will understand why children kick off their shoes as soon as their parents look away. Barefoot running on soft surfaces feels fantastic.

Judging by the overflowing shoe rack next to our back door, the barefoot-running craze has not reached the Gillen household. Though my wife and sons also run, most of the shoes are mine: cushy cruisers for long runs, mid-priced mainstays for everyday use, knobby clodhoppers for rocky trails, lightweight speedsters for racing, and quick-draining mesh-tops for spring stream crossings. You can imagine the reaction when I brought home a pair of "barefoot-running shoes." What kind of a sucker buys shoes to run barefoot? I know this may sound like the rationalization of an expensive mistake, but minimalist shoes—the term preferred by manufacturers and retailers—actually make some sense. For those who cannot always run on manicured grass or a springy track, minimalist shoes protect the feet against stones, glass, and other vagaries. Equally important, those of us who have run for decades in modern running shoes may need to ease our way into barefoot running. The Achilles and arch must handle increased stresses during barefoot running and are easily strained, causing heel and foot pain. Minimalist shoes can help bridge the gap between thick-soled traditional trainers and no shoe at all.

Though I doubt that I'll ever completely ditch running shoes, I plan to continue occasionally running without them. I'm encouraged by the sensory pleasure and improvement to my mechanics. But I've also found I need to be cautious, to build barefoot mileage slowly, and to transition slowly toward full barefoot running using minimalist shoes.

Let's turn from heels to hearts, from Achilles tendons to aortas. The beating heart cycles, filling with blood when it relaxes and pumping blood when it contracts. During contraction, the heart ejects blood into the aorta, the large artery that carries blood to muscles, kidneys, skin, and other tissues. With each heartbeat, blood pressure inside the aorta rises, pushing against the walls of the aorta and stretching them. The stretched aorta stores energy much like an inflated balloon captures some of the energy used to inflate it. And in the same way that air flows back out of a balloon when its knot is loosened, blood from the aorta surges through the body when the heart relaxes, maintaining blood flow while the heart refills.

The aorta and other arteries work much like the Achilles and other tendons. Just as tendons store kinetic energy during foot strike and return it during push-off, arteries store pressure energy during the heart's contraction and return it during relaxation. And just as tendons convert potentially jerky motions into fluid running, arteries smooth stop-and-go blood flow from the heart into more continuous flow to the body.

The physical properties of arteries resemble those of tendons in some ways and differ in others. To efficiently store and return energy, both must be resilient. But unlike tendons, which need to be stiff to transfer forces from muscle to bones, arteries must readily stretch when blood pressure rises. Thus arteries must be compliant, the opposite of stiff, and they must be extensible, able to expand without rupturing. Compliant, extensible arteries damp the high pressure generated by heart contractions, keeping blood pressure in check and lessening the risk of damage to vessel walls.

Stress on arteries escalates during exercise, amplifying the importance of compliance and extensibility. At rest, our hearts beat about 60 times per minute (min), pumping a stroke volume—the amount of blood ejected by each beat—of about 0.080 liters (L), for a total output of about 4.8 L/min. During intense exercise, the heart's output reaches 30–35 L/min because heart rate peaks at about 200 beats per minute and stroke volume rises to 0.150 L or higher.[15] The capacity of arteries to accommodate each bolus of blood resides in elastin, an abundant protein in arterial walls.

Elastin localizes with collagen in many tissues including tendons, ligaments, arteries, and skin. Although elastin matches collagen's impressive resilience, it lacks collagen's strength and stiffness. On the other hand, elastin fibers stretch up to 150 percent beyond their starting length before they snap, whereas collagen fibrils only extend about 12 percent. Elastin, in other words, is much more extensible than collagen. Envision collagen as a stiff steel spring and elastin as an extensible rubber band. When found in the same tissue, the relative abundance of collagen and elastin determines the tissue's properties. Collagen dominates in stiff tissues such as tendons, whereas elastin dominates in extensible ones such as arteries and bladder.[16]

Structural differences between collagen and elastin explain their functional differences. Like collagen, individual elastin molecules link into fibers. But unlike the linear triple helix of collagen molecules, elastin fibers are coiled like strings of cooked spaghetti.

Further, elastin fibers flap more freely because their covalent attachments are less frequent than those of collagen. Under tension, the coiled strands straighten, extending considerably before the covalent links between strands limit further stretch. The loose structure of elastin fibers fosters extensibility, allowing elastin to stretch without breaking.

Stretch a strand of spaghetti and it either snaps or stays extended. In contrast, elastin fibers snap back even when stretched to great lengths. To get a feel for this ability, grab a pair of athletic shorts and tug on the elastic waistband. Stretched to its limit, the waistband still rebounds. Likewise, resilient elastin fibers reliably recoil to their resting position, returning most of the energy used to stretch them. This resilience results from elastin's amino acid sequence, which contains many glycines, alanines, valines, and prolines. Water repels these amino acids because their side groups don't readily interact with it. These hydrophobic—"water-fearing"—amino acids tend to clump together in water, in the same way that oil separates from vinegar. Like olive oil in Italian dressing, hydrophobic amino acids gather, huddling in the interior of folded proteins, away from the water that surrounds most proteins.

You need to shake a bottle of salad dressing to mix oil and vinegar before pouring. Similarly, stretching elastin requires energy because it pulls hydrophobic amino acids away from each other into an unstable position, exposing oily amino acids to surrounding water. The hydrophobic amino acids of stretched elastin seek to reassociate, just as oil droplets in salad dressing eventually reaccumulate. Consequently, elastin tends to rebound to its original coiled position, returning some the energy that stretched it. No matter how far apart hydrophobic amino acids are pulled, their affinity for each other and aversion to water draws elastin back to its starting length.[17]

Like socks abandoned at the bottom of a gym bag, arteries get stiffer with age, losing their resilience, compliance, and extensibility. Stiff collagen increases in arteries while compliant elastin decreases, stiffening aging arterial walls. Aging also causes abnormal cross-links to form between arterial collagen fibers, further increasing stiffness. As arteries become more like rigid pipes, blood pressure increases, compounding the risk of cardiovascular diseases. Exercise ameliorates some of the age-related arterial stiffness by reversing some of aging's effects. For example, wheel running in mice counteracts arterial stiffening partly by decreasing levels of collagen.[18]

As we age, our bodies become increasingly creaky in the morning: sore muscles grumble, rusty joints gnash, and tender tendons squeal. One explanation for this morning misery might be that tendons, like arteries, become stiffer as the years go by. But in fact, tendons actually lose their stiffness as we age, impairing transfer of muscular force to bones.[19] In tendons, aging decreases collagen and increases elastin, opposite to aging's effect in arteries. Unsurprisingly, exercise increases collagen levels in tendons, reversing the effect of aging. Repeated loading of tendons during exercise triggers collagen synthesis, leading to stiffer tendons that are better able to handle high stresses.[20] Resistance exercises, such as weight lifting, stimulate collagen synthesis particularly well. Older adults benefit from resistance training partly because it enhances muscle-tendon stiffness, whereas younger athletes may need to stretch to offset the stiffness that can accompany training.[21] In summary, exercise induces changes in collagen level of both tendons and arteries that oppose the changes induced by aging.

In the second part of this chapter, we discussed elastin structure and function. Elastin's amino acid sequence and shape differ from collagen, making elastin less strong and stiff than collagen, but more extensible. We also learned that the amount of a protein matters. Tendons, rich in collagen and low in elastin, resist snapping, transmit force, and store energy because of their strength, stiffness, and resilience. Arteries, with higher levels of elastin, accommodate pulsatile blood flow because of their extensibility. Throughout this book, we'll encounter variations on this chapter's theme—that physiological function depends on both the structures and amounts of the proteins in our tissues.

Chapter 2

An Experiment of One

As I approached the vitamin store, I rehearsed my rationale for buying creatine. Though I didn't expect the clerk to ask me why a middle-aged guy like me needs an athlete's performance-enhancing supplement, I couldn't help imagining how I might respond: "It's never too late to start push-ups," or "This is for my teenage son," or "I heard this stuff improves brain power as well as muscle power," or, most honestly, "I'm doing research for a book." Needless to say, the bored clerk wasn't interested in my motivation, and I left the store without needing to make excuses.

Statements extolling creatine's performance-boosting power plastered the bottle of creatine I purchased. Each claim—that creatine improves exercise performance, that it provides energy for high-intensity workouts—was followed by an asterisk, leading to a small-print disclaimer that the assertions were not evaluated by the Food and Drug Administration (FDA). Reading the label, I began to wonder what sort of statements, approved by the FDA or otherwise, would answer my questions about creatine: Would creatine help me run faster speed workouts? Should I discourage my teenage son from taking creatine to build muscle mass? Could an elderly neighbor stay mentally sharper by supplementing with creatine?

Pages of label would be needed to account for every possible use of creatine by every category of buyer. And even if such comprehensive labels existed, potential users of creatine would still need to consider many individual factors, such as their own athletic history and goals, before deciding to use creatine. In other · words, they would need to integrate scientific research about creatine with personal information.

The value of creatine supplementation hinges on creatine's capacity to enhance the chemical energy available to our muscles. During exercise, we convert chemical energy to physical work. Chemical energy ultimately comes from the foods we eat: people transform pasta and pancakes into chest presses and pole vaults. Inside muscle cells, the same conversions happen at much smaller scales: the protein myosin burns chemical energy, generating mechanical force. But three obstacles prevent myosin from directly using food energy.

First, we typically eat only a few times a day, so food energy must be stored after meals and mobilized when needed. Second, foods are mixtures of protein, carbohydrate, and fat. To expect myosin to directly use all three nutrients would be a bit like asking a merchant to accept many forms of barter—eggs from your hens, a nifty sweater you knitted, a banjo lesson—as payment for merchandise. Perhaps possible, but definitely awkward. Finally, the nutrients we eat contain much more energy than needed by a single myosin molecule. Our digestive systems split proteins into amino acids, carbohydrates into simple sugars, and lipids into fatty acids, yet these products of digestion are still too energy rich to be used directly by muscle proteins. Even when we break pasta into individual glucose molecules, they still contain more energy than a myosin molecule can handle. Myosin is like a soda machine that demands exact change, and glucose is like a twenty-dollar bill. Thinking about it another way, imagine putting coconuts into your bird feeder—sure they're loaded with calories, but the frustrated birds would be unable to use the energy.

The difficulty of delivering energy to proteins escalates during exercise, especially in muscle cells, whose energy needs hugely increase. Muscle cells must supply their proteins with the right amount of energy at precisely the right moment. To accomplish this task, cells must transform the chemical energy of foodstuffs into usable forms. Thus, whereas Chapter 1 focuses on conversions between elastic potential energy and kinetic energy in the tendons and arteries, this chapter centers on conversions between different forms of chemical energy. These conversions, though complex, can be summarized simply: muscle cells break down sugars and fats, and harvest some of the released energy to synthesize adenosine triphosphate (ATP). For example, cells transfer the chemical energy of one glucose molecule into up to thirty-two ATP molecules.

ATP is a common currency, accepted by nearly every protein that requires energy. Further, each ATP molecule contains about the right amount of energy, the exact change that proteins like myosin need.[1] But just as we don't carry all our money in coins, our cells cannot possibly be packed with all the ATP they need. To begin with, ATP's bulky structure takes up too much space. Each of us burns about our body weight in ATP every day, much more than we can carry around. During exercise, we use ATP even faster. For example, muscles consume about 130 pounds of ATP during a two-hour run. Moreover, ATP participates in many chemical reactions and regulates myriad others, so its levels affect the rates of many cell processes. For this reason, cells go haywire when ATP concentrations fluctuate too much. Because storing ATP at high concentration is out of the question, our bodies contain less than half a pound of ATP. Each muscle only stores enough to power about one second of exercise. Thus muscle cells must constantly produce ATP, especially during exercise. Two questions arise: Where do cells get the starting materials to synthesize pounds of ATP during exercise? And how do they deal with the waste from burning so much ATP?[2]

The short answer: cells recycle. As its name implies, ATP consists of an adenosine molecule with three phosphates attached (Figure 2.1). When one phosphate breaks away, adenosine diphosphate (ADP) forms, and energy is released, as shown in this equation:[3]

Equation 1. *ATP → ADP + phosphate + energy*

Myosin and other proteins capture the energy released by ATP's breakdown and employ this energy to power their work inside cells. Rather than excreting ADP and phosphate after breaking ATP apart, cells reclaim these products to make more ATP, running the reaction in Equation 1 in reverse:

Equation 2. *ADP + phosphate + energy → ATP*

In other words, ATP resembles a rechargeable battery. We can burn many pounds of ATP during exercise because we reclaim

ADP and phosphate, reusing these molecules repeatedly. To maintain suitable ATP levels, cells must match the rates of ATP breakdown and resynthesis.

Just as battery chargers draw on electricity to revive batteries, making ATP requires energy. Phosphates tend to repel each other because they each carry a negative charge, much as the south poles of two magnets resist being pushed together. Linking phosphates together, as when a free phosphate attaches to the phosphate group of ADP, consumes considerable energy. Thus making ATP from ADP and phosphate resembles rolling a ball uphill. At the crest of a hill, a ball contains more gravitational potential energy than it does in a valley, so it takes energy to push balls up

2.1. Adenosine triphosphate (ATP) chemistry. Above left: Phosphate. Above right: Adenosine diphosphate (ADP). Below: ATP.

hills. Likewise, the product in Equation 2 (ATP) possesses more chemical energy than the reactants (ADP and phosphate), so the reaction won't flow forward without addition of energy. To sustain ATP synthesis, cells need energy that can be readily used to link ADP and phosphate.

During prolonged exercise, most of this needed energy comes from sugar and fat, as we will discuss in Chapters 5, 8, and 9. However, cranking up the systems that transfer energy from nutrients to ATP synthesis takes many seconds. A molecule called creatine phosphate bridges the gap between the time when a muscle's meager ATP stores would run dry and the time when energy from nutrients starts powering ATP synthesis. Muscle cells can synthesize ATP molecules from creatine phosphate as shown in the following equation:

Equation 3. *Creatine phosphate + ADP → creatine + ATP*

A phosphate, originally attached by a high-energy bond to creatine, transfers to an ADP molecule, forming ATP.[4] This simple transaction manufactures ATP quickly, making it a valuable producer of ATP early in exercise.

During rest, muscle cells hoard some energy as creatine phosphate, a better storage molecule than ATP. If ATP resembles nickels, then creatine phosphate resembles dimes. Creatine phosphate contains slightly more energy than ATP, but it is smaller, so it takes up less space and diffuses through cells more easily. Also cells can be packed with creatine phosphate without disturbing reaction rates that depend on ATP concentration. Despite these advantages, creatine phosphate is still a fairly bulky way to store energy, just as dimes are less convenient than bills for transporting large sums. Consequently, the creatine phosphate stored in our muscles can power only a few seconds of exercise.

Creatine supplementation aims to pack more creatine phosphate into our muscles, expanding this readily mobilized energy source. For this to work, our bodies must convert dietary creatine into creatine phosphate, a multistep process. First, the gut absorbs ingested creatine into the bloodstream. Next, muscle cells import creatine. Finally, inside muscle cells, the phosphate group of an ATP molecule transfers to creatine, forming creatine phosphate. In other words, the reaction depicted by Equation 3 runs in the opposite direction during recovery than it does during exercise, favoring formation of creatine phosphate rather than ATP.

How can a reaction run one way during rest and the other way during exercise? To help us envision the direction in which a reaction proceeds, imagine reactants and products as two rugby teams. During a rugby ruck, players from each team clash, attempting to push the other team back in order to recover the ball.[5] A team will tend to push the ruck forward if it has more players in the ruck or if its players are stronger. Similarly, both the concentration and energy of reactants and products influence the direction of a reaction.

We have already discussed an example demonstrating that chemical energy of products and reactants influences reactions. When dissolved in solutions that mimic body fluids, energy-rich ATP very slowly falls apart into ADP and phosphate, releasing energy (Equation 1). In the other direction, ATP synthesis from ADP and phosphate won't proceed on its own because it requires energy (Equation 2). All other things being equal, reactions flow from the side with the higher chemical energy to the one with the lower chemical energy, in the same way that the team with stronger players pushes a ruck forward. Creatine phosphate has higher chemical energy than ATP, so creatine phosphate tends to transfer its phosphate to ADP, driving the reaction toward the products ATP and creatine. The reaction goes this way during exercise, consuming creatine phosphate and creating ATP.

The concentrations of products and reactants also compel the creatine reaction toward ATP synthesis during exercise. High concentrations of reactants and low concentrations of products both tend to drive the reaction forward toward the products, in the same way that a ruck might be driven forward if your team adds a player or if the other team removes one. Consider the most extreme case where the reactants are abundant and products are absent—this reaction can only go forward toward products. As exercise begins, muscle creatine phosphate levels exceed those of ATP by about 500 percent, pushing the reaction toward ATP synthesis. Thus, during the early moments of exercise, the concentrations of creatine phosphate and ATP, in addition to their energy content, promote the formation of ATP from creatine phosphate.[6]

In contrast, the reaction flows in the opposite direction—toward synthesis of creatine phosphate—during recovery because concentrations of reactants and products differ compared to exercise. Once exercise ends, ATP use declines, so ATP levels increase and ADP levels fall. Meanwhile, creatine phosphate levels are low because it has been converted to creatine during exercise. ATP and creatine concentrations are higher than those of ADP and creatine phosphate, pushing the reaction toward formation of creatine phosphate. Increasing creatine concentrations through supplementation potentially tips the balance even farther in the direction of creatine phosphate synthesis.

In summary, the energy in creatine phosphate ultimately comes from our foods. While we are at rest, foods transfer energy to ATP, then ATP transfers some of that energy to creatine phosphate. Thus creatine phosphate stores small packets of food energy in a form that athletes can readily retrieve and use during exercise.

According to Equation 3, creatine phosphate and ADP combine to form creatine and ATP. But if we dissolve a pinch each of creatine phosphate and ADP in a beaker of water, very little happens. If we wait long enough, the creatine phosphate will start reacting not with ADP, but instead with water, slowly breaking down into creatine and phosphate:

Equation 4. *Creatine phosphate + water →*
creatine + phosphate + energy

No ATP will form. On the other hand, if we add an enzyme—creatine kinase—to our mixture, the beaker starts buzzing with activity. Creatine kinase and all other enzymes catalyze reactions, increasing their rates. Creatine kinase also has another role, shared by many but not all enzymes: it couples two reactions. Let's start by looking at this second function of creatine kinase.

Our beaker experiment shows that in the absence of creatine kinase, creatine phosphate gradually falls apart into creatine and phosphate, with the phosphate simply floating away rather than linking to ADP to form ATP. In this case, some of creatine phosphate's energy dissipates as heat. But in the presence of creatine kinase, creatine phosphate transfers its phosphate to ADP, generating ATP as depicted in Equation 3. By coupling the breakdown of creatine phosphate to the synthesis of ATP, creatine kinase traps energy, allowing it to be used by cells rather than lost. To say it another way, creatine kinase enables an energetically unfavorable reaction—ATP synthesis—by pairing it with a favorable reaction—creatine phosphate breakdown.

Speed matters as much in biochemistry as it does in sports. Most biochemical reactions languish without enzymes to accelerate them. We might wait hours, days, even months for a necessary product to emerge. But if it took hours to fashion ATP from creatine phosphate, the reaction would be useless. We need extra ATP while we're sprinting, not an hour later while we're showering. Enzymes increase reactions rates by a million-, a billion-, or even a trillion-fold. A molecule of creatine kinase can catalyze up to 500 transfers of phosphate to ADP per second, greatly speeding the formation of ATP. Each muscle cell synthesizes many creatine kinase enzymes, so the total ATP synthesis rate can be high enough to power high-intensity exercise, at least until creatine phosphate supplies diminish.[7]

Creatine kinase can leak from injured muscle cells into the blood. Physicians sometimes measure blood creatine kinase to assess whether patients have had heart attacks because damaged heart cells release the enzyme. Creatine kinase also spills from muscle cells into the bloodstream when an intense workout damages muscles. Athletes generally have higher levels of blood creatine kinase than couch potatoes, but show smaller increases after exercise. Still, dramatic increases in blood creatine kinase activity occur during training. In a group of college football players, creatine kinase rose from about 200 to more than 5,000 units per liter after four days of two-a-day football practices, then declined to about 1,000 units per liter after ten days of practice. These results point to substantial muscle damage during intense practice sessions, but also show that damage lessens as training continues. Chronically high blood creatine kinase might be a sign of overtraining—training so hard that performance progressively declines rather than improves—or an underlying muscle disorder.[8]

Although creatine kinase plays an essential role in muscle, its role in sperm is arguably even more crucial. That 5k race you're training for might seem important, but its significance pales in comparison to the race that begins life—the sprint of sperm to egg, a race with life or death consequences. Creatine kinase plays a slightly different role in sperm than muscle. The main role of creatine kinase in muscle is temporal—creatine phosphate forms during recovery and is consumed during exercise. In contrast, sperm creatine kinase has a spatial role, connecting the ATP production site in the sperm head to the site of ATP use in the tail.

In the sperm head, creatine kinase catalyzes the transfer of phosphate from ATP to creatine, forming creatine phosphate. This creatine phosphate diffuses to the tail, where other creatine kinase molecules catalyze the opposite reaction, forming ATP. This ATP then powers the undulations that drive sperm toward egg. Why go to all the trouble of transferring a phosphate from ATP to creatine phosphate, then back again? Couldn't ATP simply diffuse down to the tail? Recall that creatine phosphate is smaller than ATP and diffuses faster. Thus creatine phosphate moves faster from sperm head to tail than ATP would, resulting in quicker energy delivery to the sperm's tail, and perhaps making the difference between life-giving victory and the ultimate defeat.[9]

In the first half of this chapter, I've compared ATP to coins, batteries, and balls on top of hills. Let's summarize by returning to the comparison of ATP to rechargeable batteries, the example that best captures ATP's nature. People find batteries useful because they store energy, losing their charge exceedingly slowly when just sitting in a drawer. However, batteries readily donate energy when we plug them into a device. Likewise, ATP molecules are very stable in the absence of enzymes, so they are great repositories for energy. Yet when ATP plugs into the right enzymes, it readily releases some of its stored energy. Another useful feature of batteries: each type fits into numerous devices with varied functions. A radio, a flashlight, or a toy robot might harness the energy of the same battery to make sound, light, or movement. Similarly, myriad enzymes harness ATP's energy, using it to power diverse tasks like muscle contraction, assembly of cellular structures, and regulation of proteins.

Consider a champion sprinter motionless before a race. Before the starter's gun fires, we can examine the immobile runner's anatomy: how bones relate to each other and muscles attach to bones. A moment later, the runner speeds toward the finish. In motion, the body becomes more difficult to study: muscles contract and relax, bones move, limbs blur. But to understand sprinting, we need to explore runners both before and after the gun fires. Likewise, to comprehend how creatine kinase works, we must know not only its structure but also how its structure swings into action when it catalyzes reactions.

Collagen and elastin, the structural proteins discussed in Chapter 1, form long, thin filaments. In contrast, creatine kinase and many other enzymes appear globular. To better understand enzyme structures, let's compare proteins to pipe cleaners. Picture some portions of the pipe cleaner twisted into a tight spiral. Now imagine that other portions of the pipe cleaner loop back and run parallel to each other, forming flat ribbonlike features. These bends resemble two common structural motifs shared by many proteins: alpha helices and beta pleated sheets. Hydrogen bonds, the same bonds that connect collagen molecules into triple helices, glue together these motifs. Let's push the pipe cleaner analogy a bit further. Take a pipe cleaner with some helical and sheetlike regions and screw it into a complex three-dimensional shape, retaining the helices and sheets. Such folding resembles proteins' full three-dimensional shapes. Creatine kinase contains both helices and sheets folded into a globular mass (Figure 2.2). This shape—the overall form as well as the nooks and crannies—crucially influences creatine kinase's function.[10]

Creatine kinase and other enzymes accelerate and couple reactions by bringing reactants together and orienting them spatially. To appreciate the importance of this function, imagine a table

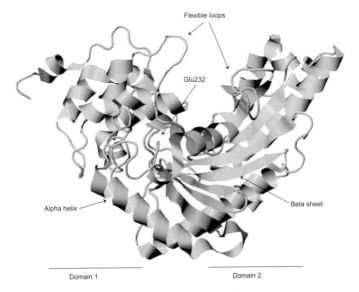

Flexible loops

Glu232

Alpha helix

Beta sheet

Domain 1   Domain 2

2.2. Creatine kinase structure. See note 10 for details about this structure.

full of puzzle pieces. To make progress, the puzzle enthusiast must find two matching pieces. Likewise, enzymes must accept only the correct reactants into their active sites—the part of the enzyme that binds reactants. Allowing the wrong reactant to bind would either prevent reactions from occurring or foster the wrong reaction. The lock-and-key model of enzyme function reflects the importance of this specificity. According to this model, reactants fit tightly into enzyme active sites in the same way that keys snuggly penetrate locks. In other words, reactants bind to enzymes because their shapes complement each other. But simply binding the right reactants isn't enough, just as finishing a puzzle requires more than simply finding pieces that match. Imagine two potentially interlocking pieces randomly moving across a table. Even if the two pieces met, they would be unlikely to make contact in a way that allows attachment. The puzzle aficionado orients pieces so that they fit together. Similarly, enzymes align reactants in a way that promotes reactions.

Creatine kinase's active site sits in a cleft between two distinct domains.[11] The ADP binding site of creatine kinase contains five positively charged arginine amino acids. These positive charges attract the negatively charged phosphate groups of ADP, resulting in an active site that precisely binds ADP but not other compounds. Likewise, creatine phosphate fits securely into its binding site on creatine kinase. Thus creatine phosphate and ADP react with each other rather than with whatever molecule happens to be floating by. In this way, creatine kinase encourages transfer of phosphate from creatine phosphate to ADP and discourages other promiscuous activity.

Our discussion so far portrays creatine kinase molecules as static, like a sprinter in the starting blocks. But creatine kinase does not simply passively bind reactants. Instead, it actively facilitates their interactions. Just as a runner's leg muscles start contracting and relaxing when the race starts, creatine kinase also springs to life when the gun goes off.

The induced-fit model of enzyme function depicts enzymes as malleable rather than rigid. In this model, reactants morph enzymes, causing changes in the enzyme's shape. Or to put it another way, enzymes actively embrace reactants. The two domains of creatine kinase each contain flexible loops with charged amino acids—positive on one loop and negative on the other. As ADP and creatine phosphate bind, creatine kinase morphs. This contortion brings opposing charges on separate loops together, and their attraction latches the loops over the active site. Amino acids near the reactants move too. These movements create a contained binding pocket, align reactants so that they react more readily, and bring parts of the enzyme into intimate contact with the reactants.[12]

At the start of a sprint, the movements of muscle creatine kinase molecules resemble those of calf muscles. While the calf contracts and relaxes, creatine kinase clasps ADP and creatine phosphate into its active site, and then opens to release ATP and creatine. But one difference is the speed of these cycles. Creatine kinase catalyzes hundreds of reactions in the same time it takes a calf muscle to contract and relax. Thus creatine kinase cycles through its conformations much more quickly than muscles contract and relax.

Although the induced-fit model more accurately represents the dynamic activity of enzymes than the lock–and–key model, creatine kinase and other enzymes do more than simply cuddle reactants. Reactants typically don't transition directly into products;

instead, they progress through transition states—short-lived structures that readily convert back to reactants or proceed to products. Enzymes pave the path from reactant to product by assisting formation of these intermediate steps. For example, when ADP and creatine phosphate bind to creatine kinase, the enzyme squashes and stretches these reactants into a transition state that looks more like the products: ATP and creatine. As a result, the products form much faster than they would without the enzyme's intervention.[13]

Enzymes encourage transition states by assembling, aligning, and morphing reactants. Some enzymes also form short-lived bonds with reactants, temporarily donate a small piece of themselves to reactants, or briefly accept a portion of a reactant. In summary, enzymes do more than stand on the sidelines urging reactions along. Rather, they jump into reactions, contorting themselves as well as their reactants. Enzymes thereby actively engage with reactants, prodding them to convert to products.

Often, scientists can identify the specific amino acids involved in these interactions between enzymes and reactants. In one approach—site-directed mutagenesis—they synthesize proteins with changes at single amino acid positions. For example, when scientists postulated a key role for a glutamine amino acid at position 232 of creatine kinase (Glu232), they created creatine kinase enzymes with different amino acids substituted for Glu232, and then they assessed the enzymes' activities. When the substituted amino acid was structurally similar to glutamine, enzyme activity decreased about 500-fold. However, when a very dissimilar amino acid replaced Glu232, scientists observed decreases up to 100,000-fold. Substitutions at a different site produced substantially smaller decreases in activity. This work demonstrates that

Glu232 is important for creatine kinase function. And more generally, studies like these show that small changes in amino acid sequence can have huge impacts on enzymatic function.[14]

In summary, enzymes resemble referees. Officials of basketball, football, and other sports organize the game and ensure that players adhere to the rules, just as enzymes manage and facilitate reactions. Moreover, referees closely interface with the game. They run the court or field, they respond to player's movements, they handle the ball, and in some sports (think hockey) they even interpose themselves in tussles between opponents. Similarly, creatine kinase and other enzymes participate physically and chemically in reactions. They change shape, make bonds, temporarily donate parts of themselves, and receive fragments of reactants.

On the other hand, referees—at least in the ideal world—are unaffected by the outcomes of games. Their happiness or wealth should not depend on which team wins. Likewise, enzymes emerge unscathed at the end of a reaction, returning to their original forms, ready to catalyze another reaction. Finally, good officials do not influence the outcome of games by playing favorites, in the same way that enzymes do not establish which way a reaction proceeds. Creatine kinase can convert ADP and creatine phosphate to ATP and creatine, or catalyze the reverse reaction of ATP and creatine to ADP and creatine phosphate. The concentrations and energy content of these molecules, not the whims of creatine kinase, determine the reaction's direction.

The biochemistry of ATP, creatine, and creatine kinase offers a straightforward rationale for creatine supplementation—ingesting creatine should increase muscle creatine phosphate and improve short-term, high-intensity exercise performance. But studies of creatine paint a more complex picture. Some work shows that creatine loading leads to big increases in muscle creatine phosphate; other work shows smaller effects. Some studies find that creatine improves exercise performance; others find no effect. Sometimes performance gains are quite small; in other cases they are substantial. In fact, the question "Does creatine supplementation improve performance?" may be most accurately answered "It depends." Although many factors contribute, the effectiveness of creatine supplementation depends mainly on the type of exercise and subject population.

Most studies demonstrate that creatine loading increases performance during short-term, heavy exercise. The largest benefits occur during repeated high-intensity exercise bouts, with improvement typically in the 5–20 percent range. For example, in one study volunteers performed three sets of maximal knee extensions—quick, explosive contractions. Creatine supplementation resulted in a 10–23 percent increase in force production, with the biggest effects at the beginning of each set.[15] Many other studies show substantial increases in maximal strength. Creatine loading also results in benefits during single bouts of continuous exercise, like thirty-second bursts of maximal cycling effort, but these effects tend to be smaller, with improvements typically around 1–5 percent. Most studies show no benefit of creatine supplementation for endurance exercise.[16]

So the benefits of creatine loading mainly apply during short, intense exercise, consistent with the role of creatine phosphate as a readily available supply of energy that can power a few seconds' worth of exercise. Yet despite its rather specific effect on exercise performance, many additional effects have been attributed to creatine intake. According to recent studies, creatine increases muscle mass, improves memory, and helps cells regulate calcium. Creatine also apparently protects cells; it serves as an antioxidant, shielding cells from the harmful effects of oxygen, and as a neuroprotectant, defending nerve cells against damage. Although still in their infancy, clinical studies on creatine hold promise for the treatment of many conditions, including epilepsy, traumatic brain injuries, and Huntington's, Alzheimer's, and Parkinson's diseases. And though anecdotal reports blame creatine for causing liver, kidney, and gastrointestinal damage, most studies uncover little evidence for these side effects during creatine loading at recommended dosages. Much more work needs to be done—on creatine's clinical effectiveness, the effect of ingesting creatine in combination with other substances, and on optimal dosages—but an emerging consensus suggests that creatine has broad potential in treatment of human disorders.[17]

The efficacy of creatine for both exercise performance and disease treatment may depend upon subject populations. In the elderly, the effects of creatine on exercise performance are small, though benefits have been demonstrated. Effects of creatine, particularly gains in muscle mass, are sometimes bigger in men than in women. Diet also matters. The main dietary sources of creatine are meat and fish, so creatine supplementation may be particularly beneficial for vegetarians. Also, caffeine may block the creatine effect, so chronic coffee drinkers may see less benefit. And finally, some individuals are low-responders, showing very small or no effect of creatine supplementation. Consequently, we see that the efficacy of creatine supplementation depends on many factors, and decisions about creatine loading must be tailored to specific individuals and goals.[18]

The running guru Dr. George Sheehan famously wrote, "Each of us is an experiment of one."[19] This maxim applies to decisions about taking supplements as well as it does to training and nutrition. In the end, we all have different bodies, goals, schedules, attitudes, and risk tolerance. What works for me may not work for you. Dr. Sheehan was not suggesting that we treat our bodies lightly, that we ignore established scientific evidence, that we rush thoughtlessly into experiments on ourselves, or that we make ourselves a testing ground for every new trend. Neither am I. Going on a diet, taking supplements, altering sleep patterns, and trying a new training program are all serious choices, ones that can harm or benefit both health and athletic performance. Yet the best we can do is consult teammates, coaches, health care professionals, and the scientific literature, and then choose our course and see what happens. Eventually, we need to run the experiment.

I ran an "experiment of one" with the creatine I purchased, taking the pills for a month or so while I tried to build some upper-body strength with push-ups. I'm a vegetarian, so I figured that my muscle creatine might be lower than most people's. If extra creatine helped, the effects were small. Perhaps my dose was too low, or my push-up program too moderate. Or perhaps my skinny runner's body resists muscle building, even when helped along by creatine. Maybe I would experience noticeable benefits of creatine supplementation on my running, particularly during high-intensity speed workouts.

We've seen in this chapter that creatine phosphate serves as an important fuel during intense exercise and that ingesting creatine can elevate creatine phosphate levels in muscles. Yet accurately predicting whether a particular athlete will benefit from taking creatine remains a challenge because the effects of creatine supplementation depend on many factors. To make good decisions about its use, athletes need to consider creatine's biological role in light of their own physiology, training goals, risk tolerance, and previous experiences with diet, supplements, and training. No simple, one-size-fits-all rule suffices. For me, complexity like this makes discussions about exercise performance fascinating.

Chapter 3

The Gene for Gold Medals

At nine years old, I toed the line of my first "marathon," planning twenty-six laps around the block. The other kids stopped after a block or two, but I kept plodding, emulating the Olympic marathoners I had just watched. Suffering soon set in. The block turned out to be longer than I thought, the day became progressively hotter, and my friends wandered off to more sensible activities. But I found joy underneath the misery of those lonely laps, and though I quit long before the twenty-sixth loop, I was hooked on distance running.

In the twenty-five years that passed before I ran an actual marathon, I increasingly envisioned myself as an endurance athlete. On the tennis courts, my ground strokes were weak, but I still hustled in the third set. On the soccer field, I was slower than teammates, but less tired in the second half. While others loathed the laps we ran after practice, I secretly enjoyed them. When friends hit the weight room, I pounded the pavement.

Did my passion for distance doom me to athletic mediocrity? Maybe I missed my calling as a shot-putter or sprinter? Suppose I had been spellbound by the hundred-meter dash or javelin instead of the marathon? What if I had set up hurdles rather than a distance course? For me, and other athletes who rank enjoyment over achievement, the answers to such questions don't matter much. But for those with Olympic aspirations, or even hopes of a college scholarship, choosing the right sport might be crucial.

My decision to emphasize endurance was more intuitive than scientific. Today, genetic testing promises to eliminate some of the guesswork from these decisions. Genetic testing hinges on our ability to connect variations in genes to disease susceptibility, athletic prowess, and other physiological traits. To make these connections at the molecular level, we attempt to link DNA, the genetic material, to proteins, whose structures ultimately produce function.

The ACTN3 gene instructs cells to make actinin-3, one member of the actinin protein family.[1] ACTN3 has been called a "gene for speed," taking its place alongside other "gene for" genes: the gene for sleep, the gene for left-handedness, the gene for speech, and, perhaps someday, the gene for gold medals. The hypothetical (and admittedly slightly ridiculous) gold-medal gene highlights the shortcomings of talking about the gene for this or that. Such language conceals the twisted paths linking genes, proteins, and physiology. First, most traits are multigenic—multiple genes contribute to the physiological outcome—so *one of the genes for* is often more accurate than *the gene for.*

Additionally, genes don't always map one-to-one with functions. The "one gene, one enzyme" hypothesis influenced the emerging science of molecular biology in the 1940s and 1950s. The essence of this hypothesis—that a gene controls the synthesis or activity of each enzyme—has been spectacularly confirmed. We now know that discrete stretches of DNA contain genes that direct the synthesis of specific proteins with recognizable functions. On the other hand, the simplest interpretation of the hypothesis—that one gene makes one protein which performs one task—oversimplifies a complex molecular world. For example, in Chapter 11 we discuss a process called splicing, whereby one gene produces more than one protein. Also, some enzymes—such as the muscle protein myosin—require multiple subunits, so it sometimes takes more than one gene to make a functional molecule. Furthermore, the relationship between proteins and functions is also complicated. Many proteins—including actinin—execute multiple duties. The converse also holds: certain functions can be performed by more than one protein. For example, closely related genes sometimes make sibling proteins, called paralogs, with very similar activities. Actinin-3 is one of four actinin paralogs in the human genome.[2]

One of the first genetic tests for athletic prowess assessed ACTN3 varients. Though researchers have moved beyond single-gene tests, the ACTN3 test offers us an oppotunity to probe the complex relationship between genes and proteins.

Actinin proteins participate in a network of struts, strands, bars, and beams—the cytoskeleton—that crisscrosses inside every human cell, including muscles. The cytoskeleton's functions resemble those of our bone skeleton. Bones support the body, transforming formless flesh into distinct human form. In the same way, the cytoskeleton buttresses cells, helping them contort into fabulous shapes that resemble pyramids, columns, starbursts, and pancakes. Bones also mediate movement, transferring muscular force into physical actions. Likewise, the cytoskeleton underpins cellular motion. Many cells—not just muscles—generate force and motion. All cells grow and divide, changing their shapes in the process. Some cells creep along surfaces. Some propel themselves with whiplike flagella. Some attach to other cells and then stretch and pull. Moreover, transportation networks inside cells ferry materials from one place to another. These movements of cells and their components rely on the cytoskeleton.[3]

In the cytoskeleton, actinin associates with actin, a protein with a similar name but very different function. Actin plays a leading role in muscles, yet it is found everywhere, not only in all human cells but in cells of almost all organisms. Actin is enormously abundant in muscle cells, about 10 percent of total protein, and merely plentiful in other cells, where it is 1–5 percent of total protein.[4] Amoeba and plants, though they lack muscle, contain actin that is 80–90 percent identical in amino acid sequence to animal actin. In non-muscle cells, actin contributes to cell shape, intracellular cargo movement, amoeboid creeping of cells, and cell division. These roles depend on actin's capacity to form long chains. Under the right conditions, hundreds of individual actins—each a moderate-sized globular protein—assemble into lengthy filaments. A mesh of these actin filaments lies underneath the cell membrane.[5]

Actin supports motion in two ways. In some cases, rapid assembly of actin filaments drives movement. For example, formation of filaments at the edge of a cell can push outward on the membrane, projecting a pseudopod (literally "fake foot") and thereby propelling amoeboid movement. In other cases, stable actin filaments serve as a scaffold for activity of myosin proteins. Myosin molecules sometimes walk along actin filaments, like trains transporting goods along railroad tracks. And in kidney, liver, muscle, and many other cells, strands of myosin tug on actin filaments and produce force.

If actin and myosin collaborate to produce force, then what does actinin do? Two elongated actinin molecules align head to tail, forming dumbbell-shaped molecules with actin-binding sites at both ends.[6] These short actinin dumbbells cross-link long actin filaments, strengthening the cytoskeleton. Thus actin filaments resemble girders, connected by actinin linkers. Moreover, actinins bind not only actin but other proteins too, thereby linking actin filaments to other cellular structures. Attachments of actinin to proteins that span the cell membrane, for example, allow the actin cytoskeleton to pull on the membrane, affecting cell shape. In Chapters 1 and 2, we looked at proteins either working on their own or associating with other proteins of the same type, as in collagen fibrils. In this chapter, we begin to see proteins of different types working together, rubbing shoulders with each other in multiprotein complexes.

Cytoskeletal structures called sarcomeres pack muscle cells. In sarcomeres, actin and actinin play roles very similar to the ones they play in the cytoskeleton of other cells. Thus the distinctive talent of muscles—the production of force—emerges not from an entirely novel set of proteins, but rather from refinements of cytoskeletal components found across animal cells. This example points to a theme of this chapter: breathtaking variation often stems from deep underlying commonalities. In short, diversity arises from unity.[7]

Muscles exemplify the ying-yang of unity and diversity in biology. Although all muscles rely on the same fundamental mechanisms, they exhibit an astonishing diversity of properties. Variation exists across species, in different muscle types within the same species, and among individuals of the same species. Muscles from different animal species show stupendous diversity. Superfast rattlesnake tail muscles contract and relax almost a hundred times per second. Fantastically fatigue-resistant retractor muscles of clams, oysters, and mussels contract for hours with very low energy use, holding the two shells of these animals tightly together. And certain salamanders' remarkably powerful tongue muscles contract with so much force that they fire the tongue almost a full body length in less than twenty milliseconds. Don't try any of these tricks yourself—your muscles simply are not up to the task.[8]

Although human muscles do not quite reach the extremes of these animal examples, they still show a wide range of properties. Our muscles group into three classes: skeletal muscle moves bones; cardiac muscle pumps blood; and smooth muscle constricts hollow structures such as guts, airways, and blood vessels. Though skeletal muscle plays the most obvious role during exercise, smooth and cardiac muscle contribute too. Both help control internal conditions—such as blood pressure and oxygen levels—in the face of the stresses that accompany exercise. For example, arterial smooth muscle maintains a constant tone that helps regulate blood pressure, and cardiac muscle drives oxygen-rich blood to tissues.

Cardiac, skeletal, and smooth muscles require different functional properties. Cardiac muscles have high endurance because they must cyclically contract and relax for decades without rest breaks. And cardiac muscle cells must activate in synchrony because an effective heartbeat demands coordinated contraction.

Though smooth muscle typically makes slow, sustained contractions, it must be effective across a wide range of lengths because the hollow organs it surrounds swell and shrink. Skeletal muscle cells, called fibers, vary greatly in their properties, sorting into three different types. Type 1 fibers—slow twitch—resist fatigue but are not very powerful. In contrast, type 2X fibers—fast twitch—fatigue quickly but are quite powerful. Type 2A fibers are intermediate between type 1 and type 2X.[9] Though each human skeletal muscle contains a mixture of fiber types, the ratios vary across muscles. Consider the calf, which contains two distinct muscles, the soleus and the gastrocnemius. Type 1 fibers dominate in the soleus muscle, whereas type 2X fibers dominate in the gastrocnemius. So the soleus excels at sustained contractions, such as the ones required to stand at attention, whereas the gastrocnemius excels at powerful contractions, like the ones needed to clear a hurdle.

Muscles also vary among individual humans. Tortoises like me are rich in slow-twitch fibers, whereas the hares that dart past us tortoises are rich in fast-twitch fibers. Training can alter ratios of slow- to fast-twitch fibers, but substantial differences exist among individuals prior to training. If scientists could readily determine whether slow- or fast-twitch fibers dominate in a person's muscles, they might be able to recommend certain training programs or encourage participation in certain athletic events. Folks loaded with slow-twitch fibers might be urged to run cross-country, whereas those loaded with fast-twitch fibers might be encouraged to compete in sprinting events. Fiber types can be assessed in muscle biopsies, small samples of muscle taken via a hollow needle. But most athletes shy away from having their muscles punctured. Genetic tests promise a less invasive means of sorting people into tortoises and hares. The ACTN3 test takes a step in this direction.

If we properly prepare skeletal or cardiac muscles and view them through a microscope, an astonishing pattern appears. Dark and light bands stripe each muscle cell, dividing the cells into sarcomeres—the basic unit of muscle function. Each sarcomere produces only tiny forces and shortens miniscule distances, but an astounding number of sarcomeres—about 200 million per muscle cell—cooperate to produce contractions. Sarcomeres, in turn, rely on the collaboration of millions of individual protein molecules. The ordered banding pattern seen in skeletal and cardiac muscle reflects the strict organization of muscle proteins into sarcomeres.[10]

Myosin and actin—the proteins that cooperate to produce force—form the main structure of sarcomeres (Figure 3.1). Myosin proteins, arranged into strands called thick filaments, occupy the center of each sarcomere. Actin proteins, assembled into thin filaments, interdigitate with thick filaments. To model these filaments, lay two pens parallel on a table. Now, spread the first three fingers on each of your hands, place them flat on the table, and slide them between the pens so that pens and fingers alternate. Refrain from pushing your hands together, leaving gaps between the tips of your fingers. The pens represent thick filaments and your fingers represent thin filaments. Though pens and fingers offer a useful model of muscle proteins, actual sarcomeres have many more than two thick filaments. Also, sarcomeres are three-dimensional, so six thin filaments surround each thick filament, rather than the two depicted in our two-dimensional model.

The vertical bands discernible on micrographs—images taken through microscopes—run perpendicular to the thick and thin filaments. The protein myomesin attaches to parallel thick filaments, forming a visible band that bisects the sarcomere called the M line. To model the M line, lay a crayon across the pens of your sarcomere model. At the outer boundaries of the sarcomere, actinin molecules form thin dark stripes called Z lines. Envision the webbing between your fingers as these Z lines. Actinin molecules cross-link thin filaments that protrude from the Z line. Actinin's role in the sarcomere mirrors its role in the cytoskeleton. It binds nearby actin molecules, stabilizing their arrangement. But compared to the cytoskeleton, where fila-

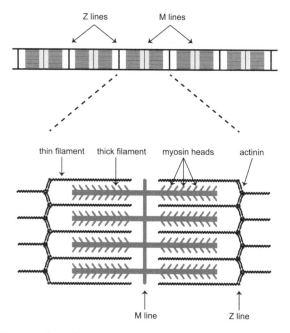

3.1. Diagram of muscle sarcomere. Above: Banding pattern of sarcomeres arranged in series. Below: Structure of one sarcomere. Not drawn to scale.

ments run in many directions, the sarcomere is highly ordered, with thick and thin filaments positioned perpendicular to the M and Z lines. So in sarcomeres, actinin holds actin in a precise orientation.

Let's return to your pen–and–finger model. Adjust your hands to create a space between the pens and the webbing at the base of your fingers so that your fingers alone occupy the spaces nearest your palms. In the sarcomere, the corresponding regions contain thin filaments but no thick filaments. These areas appear as the lightest parts of muscle micrographs. Closer to the center, your fingers overlap with pens, corresponding to places where thin and thick filaments overlap, the darkest regions on micrographs. Finally, only pens occupy the center of your model, corresponding to the portion of the sarcomere with only thick filaments. In this area the shading on micrographs is intermediate.

Now, slide your fingers back and forth. Notice that as the distance between your hands changes, the amount of overlap between pens and fingers also changes. In the 1950s, scientists observed that the length of skeletal muscles affects their banding pattern. For example, the lighter bands at the center and ends of sarcomeres become thinner as muscles shorten. This thinning happens because the regions containing only thin or thick filaments shrink, just as the space without overlap between your fingers and pens lessens when you slide your hands together. These observations led to the sliding filament theory of muscle contraction—the idea that thick filaments attach to thin filaments and tug them toward the sarcomere's center, thereby generating force. In Chapter 4, we will explore the interactions between the myosin proteins of thick filaments and the actin proteins of thin filaments that power force production.

In summary, proteins of different sizes and shapes assemble into a highly-ordered, multiprotein complex in sarcomeres. Though we have discussed roles for many proteins—actin, myosin, actinin, and myomesin—other proteins are also involved. The sarcomere proteins troponin and tropomyosin regulate muscle, connecting signals from nerves to contractions of muscle fibers, as described in Chapter 6. Another sarcomere protein—calsarcin—helps regulate muscle properties, as we will see later in this chapter. Actinin, our focus at the present, plays a key role in sarcomeres, binding actin molecules and thereby holding thin filaments in their parallel relationship to thick filaments. We turn now from the actinin protein to the gene that encodes it.

The ACTN3 genetic test differentiates two alleles—versions—of the ACTN3 gene. Alleles resemble different editions of the same book, nearly identical except for a few typos or corrections. Yet the small differences between alleles sometimes result in substantial functional consequences. One ACTN3 allele favors excellence at power sports such as weight lifting and sprinting. The other allele might enhance success at endurance sports such as distance running and cycling. To assess which ACTN3 alleles an athlete possesses, scientists first collect the athlete's cells, often from the inside of the cheek. Next, they isolate DNA, separating it from other cellular components. Finally, they evaluate the variable portion of the ACTN3 gene to determine which alleles are present.

Alleles differ in their deoxyribonucleic acid (DNA) sequences, the linear order of nucleotides linked together in DNA. Nucleotides share similarities with the amino acids that join into proteins. Like the twenty different amino acids, the four different nucleotides that form DNA are variations on a basic structure

(Figure 3.2). Each nucleotide possesses the same sugar and phosphate groups. And much as bonds between the invariant groups of amino acids link them into proteins, bonds between the sugar of one nucleotide and the phosphate of another connect nucleotides into DNA strands. Individual nucleotides derive their distinct properties—and their names—from the different base groups they contain: thymine (T), adenine (A), cytosine (C), or guanine (G). These base groups protrude from the sugar-phosphate backbone. A single nucleotide difference—T versus C—differentiates the two ACTN3 alleles.

3.2. Deoxyribonucleic acid (DNA) structure. Above: The four bases of DNA. Below: Portion of a DNA strand depicting the sugar-phosphate backbone and the variable base groups.

The order of linked nucleotides in DNA conveys information, much like the sequence of letters in a book. But the DNA alphabet contains only four letters, many fewer than English and other languages. With more options at each position, English packs information into shorter sequences than DNA. Thus DNA requires many nucleotides to store our genetic information. The longest DNA chains in humans contain almost 300 million nucleotides and are almost ten centimeters (cm) long when stretched straight.[11] Of course, elongated DNA strands would never fit into cells, so our DNA coils tightly around protein molecules, forming chromosomes. Our entire genome contains about three billion nucleotides across twenty-three pairs of chromosomes.

Genes are distinct portions of chromosomes that contain instructions for making a particular protein. In other words, genes "code for" proteins. ACTN3 encompasses a small stretch along human chromosome 11 and codes for the protein actinin-3. To represent the amino acids of a protein, nucleotides of genes string together into three-letter words. Nucleotide triplets form sixty-four different combinations (4 × 4 × 4), so in some cases multiple triplets code for the same amino acid. Though ATG is the only triplet that encodes the amino acid methionine, the triplets CCC, CCT, CCA, and CCG all signify proline, one of the key amino acids in collagen (Chapter 1). The triplet code also contains one type of punctuation mark—TAA, TAG, and TGA triplets mean "stop," signaling the end of a protein. Cells contain machinery that decodes DNA, reading a gene's triplet code and linking the correct amino acids into a protein. For example, the coding region of ACTN3 begins with the nucleotides ATG ATG ATG, so the first three amino acids of actinin-3 are methionine.

The triplet code is universal, shared not only by record-holding sprinter Usain Bolt and marathoner Paula Radcliffe but also by microbes, mushrooms, zinnias, and zebras. The nucleotide combinations representing proline are identical across every living creature, as are the combinations for methionine, alanine, and all the other amino acids. Thus astounding physiological and morphological variation comes from the same genetic language, much as authors have written an extraordinary range of works in English— from Macbeth and Middlemarch to the Hardy Boys and Harry Potter. The genetic code represents perhaps the most impressive example of diversity flowing from unity.

How do different alleles of ACTN3 and other genes arise? When cells divide, they replicate their chromosomes, copying the genome with remarkably high fidelity. Only one mistake typically occurs per billion nucleotides. When such errors happen in liver, muscle, brain, and most other cells, they do not pass to offspring. In contrast, mistakes made in copying DNA during production of gametes—sperm and eggs—by testes and ovaries pass to offspring if the affected gamete gets together with its complement to make an embryo. Alleles result from these inherited mutations. Most mutations are harmless. Regions that code for genes encompass only about 2 percent of the human genome, so most mutations miss coding regions.[12] And even mutations within coding regions sometimes leave proteins unscathed. Occasionally, as is the case for the ACTN3 alleles, mutations seriously alter protein function. So mutations have variable effects on proteins, just as typos have variable effects on the meaning of sentences.

Mistakes that insert or delete nucleotides often cause severe problems. Consider the sentence "Everything the king touched turned to gold." Now, delete the first "E," shift the other letters to fill the open space, and keep the word lengths the same: "Verythingt hek ingt ouchedt urnedt og old." The sentence loses its original meaning—in fact, it loses all meaning. Likewise, insertions and deletions shift the "reading frame" of genes, often changing every amino acid that follows. For example, if we delete the first nucleotide of the ACTN3 gene, we go from ATG ATG ATG to TGA TGA TG. Now, the first two triplets are TGA—stop signals—a change that clearly alters the resulting protein. Fortunately, insertions and deletions occur very rarely.

The ACTN3 alleles stem from a point mutation in the ACTN3 gene. More common than insertions and deletions, point mutations occur when one nucleotide substitutes for another. Return to the sentence "Everything the king touched turned to gold" and replace an "e" for the "u" in "turned." The meaning doesn't change. Likewise, some point mutations are silent; they have no effect on protein sequences. For example, mutations that switch CCC to CCG, CCA, or CCT are silent because all these triplets code for proline. On the other hand, replace the "g" in gold with an "m" and the king's power becomes considerably less enviable. Similarly, some point mutations cause amino acid substitutions. For example, if the triplet CCC mutates to GCC, alanine replaces proline. Single amino acid changes can have huge or subtle impacts on function. Recall from Chapter 2 that mutations of a key amino acid in creatine kinase decrease the enzyme's activity by 100,000-fold, whereas mutations of other amino acids have small effects.

The ACTN3 genetic test detects a special kind of point mutation called a nonsense mutation. Nonsense mutations insert a stop signal into a gene's sequence, prematurely terminating protein synthesis. As a result, the gene codes for truncated proteins. In the ACTN3 nonsense mutation, a change from C to T converts an arginine (CGA) to a stop (TGA) at position 577 of the amino acid sequence. The 577R allele—with arginine (R) in position 577—codes for normal, full-length actinin-3. In contrast, the 577X allele—with a stop (X) at position 577—produces pruned proteins that fail to function.

Yet, even though it results in broken actinin-3 protein, the 577X allele appears commonly in people, with frequencies above 50 percent in some human populations. On the other hand, 577X is less common among elite power athletes, such as Olympic sprinters and lifters, suggesting that loss of actinin-3 impairs performance in such events. Finally, some evidence suggests that 577X may be more prevalent among top athletes in distance events, meaning that 577X might be beneficial to endurance.[13]

How do many people thrive with broken actinin-3? One reason is that our cells contain two copies of the ACTN3 gene and almost all other genes—one from our biological mother, the other from our biological father. Both copies usually produce proteins. For each gene, we either have two different alleles (the heterozygous condition) or two copies of the same allele (the homozygous condition). In the heterozygous condition, cells make two versions of the same protein. People who are heterozygous for ACTN3 carry one 577R and one 577X allele. They suffer no serious effects, probably because the 577R allele produces enough normal actinin-3. But approximately 16 percent of people are homozygous for 577X and therefore manufacture no functional actinin-3 protein. Amazingly, complete loss of this protein confers no obvious harm. In fact, the frequency of the 577X allele is increasing in some populations, suggesting that it may carry benefits.[14]

To understand why 577X homozygotes experience only minor consequences, we need to consider the paralogs of actinin. While alleles are variants of the same gene, paralogs are separate genes that encode very similar proteins. If alleles are like different editions of the same Nancy Drew mystery, one with a few misprints, then paralogs are like different titles in the Nancy Drew series. *The Hidden Staircase* and *The Bungalow Mystery* are closely related in content, structure, and style, but dissimilar enough to merit separate titles. Actinin paralogs resemble each other to a much greater degree than they resemble collagen, myosin, and other proteins. If actinin paralogs are separate volumes in the Nancy Drew series, then actinin and collagen differ in the same way as the Nancy Drew and Harry Potter series differ. No one is going to mistake *Harry Potter and the Sorcerer's Stone* for a Nancy Drew mystery, and no one who looks closely could mistake actinin for collagen.

Paralogs originate in duplications of entire genomes or individual genes—both very rare events. Duplicated genes slowly diverge from each other through mutations that arise during DNA copying, resulting in proteins with slightly altered structure and function. So paralogs exemplify the tension between unity and diversity. Paralogs are highly similar, yet divergent enough to play distinct roles in cells.

Four genes code for actinin paralogs with distinct localizations in the human body. Two actinin paralogs, actinin-1 and actinin-4, contribute to the cytoskeleton of all cells. In contrast, actinin-2 interacts with actin in the sarcomeres of skeletal and cardiac muscle cells, but is absent from most other cells. Actinin-3 has an even more restricted territory—sarcomeres of fast-twitch (type 2X) skeletal muscle fibers. Smooth muscles, which lack sarcomeres, also lack actinin-2 and actinin-3 expression.[15]

The limited expression pattern of actinin-3 may partly account for the subtle effects of ACTN3-577X mutations: only activities that use fast-twitch skeletal muscle, such as sprinting and power-lifting, should be affected. Studies of elite athletes support this explanation. Very few athletes with nonfunctioning actinin-3 excel in power events. A study of Australian athletes found only 6 percent of elite sprinters to be 577X homozygotes, compared to 18 percent in a control population. Of thirty-five female Olympic sprinters in the same study, none were 577X homozygotes. On the other hand, the 577X allele was present in elite endurance athletes at the same or even higher levels than controls, so the 577X mutation seems to impair only high-intensity exercise.[16]

The 577X mutation might also produce understated effects because other paralogs offset the loss of actinin-3. Actinin-2 is the obvious candidate—it shares 79 percent amino acid identity with actinin-3, and both proteins reside in skeletal muscle. To explore the effects of actinin-3 loss, researchers engineered a line of mice that do not produce it. These actinin-3 knockout mice make extra actinin-2, suggesting that mice overexpress actinin-2 to compensate for actinin-3 loss. The mice without actinin-3 lived healthy lives, so actinin-2 probably successfully replaced actinin-3. Presumably, actinin-2 compensates for actinin-3 in humans too.[17]

Nonetheless, actinin-2 and actinin-3 do not perform identical roles. Mice and humans lacking actinin-3 apparently suffer impaired power. Why is actinin-3 required for optimal power? The strongest muscle contractions tug hard, straining the proteins of the sarcomere. As a tie between actin molecules, actinin bears some of this tension. Actinin-3, specialized for fast-twitch muscle, may be more adept at handling these forces than actinin-2. So although actinin-2 may adequately replace actinin-3 during ordinary circumstances, it might not be as effective during high-intensity exercise. Hence the basis of the ACTN3 genetic test: athletes need at least one copy of ACTN3 577R for elite performance in power and sprinting events because actinin-3 is specialized for fast-twitch muscle.

Although very strong evidence links ACTN3 577R to success in power events, less convincing evidence suggests that ACTN 577X provides benefits in endurance events. In human studies, 577X turns up with only slightly greater frequency in elite endurance athletes than in the general population. Yet a small boost in endurance capacity might explain why the 577X allele persists— and perhaps even is increasing—in human populations despite its negative consequences for muscular power. Although it is easy to see how loss of actinin-3 could harm performance in power events, it is tougher to imagine how loss of any protein could improve endurance.

A possible explanation for how actinin-3 loss benefits endurance hinges on the possibility that actinin has multiple roles. In other words, the tendency of ACTN3-577X homozygotes to have increased endurance may stem from interactions of actinin with proteins other than actin. In addition to actin, actinin binds calsarcin (another sarcomere protein), fructose bisphosphatase (a metabolic enzyme), the acetylcholine receptor (a membrane protein that we discuss in Chapter 6), and many other proteins. Substituting actinin-2 for actinin-3 could alter interactions with these proteins, thereby shifting muscle properties. Let's look at the consequences of actinin interactions with calsarcin and fructose bisphosphatase.

Calsarcin influences muscle differentiation, affecting whether new muscle cells grow into fibers that specialize in endurance or power. Knockout of calsarcin activates pathways that shift muscle fibers toward the slow-twitch type, thus resulting in enhanced endurance. Actinin-2 and actinin-3 may exert different influences on calsarcin when they bind it, so muscle fiber type may depend partly on which actinin paralog is present. Fructose bisphosphatase catalyzes a reaction in the pathway that produces glycogen, an important fuel source during endurance activities (Chapter 9), and also serves as a sensor, turning glycogen synthesis on or off depending on the circumstances. Replacing actinin-3 with actinin-2 probably modifies fructose bisphosphatase's behavior, leading to greater glycogen accumulation and enhancing endurance.[18]

The following is a potential mechanism for increased endurance in ACTN3-577X homozygotes: actinin-3 loss triggers increased actinin-2 expression, altering the behavior of numerous proteins that associate with actinin. As a result, signaling and metabolic pathways change in ways that encourage a shift toward muscle with slow-twitch properties. Animal studies offer some support for this idea. Both endurance training of rats and actinin-3 knockout in mice increase actinin-2 levels and shift muscle cells toward aerobic metabolism—a hallmark of fatigue-resistant slow-twitch muscle fibers. Though much work needs to be done to confirm this hypothesis, it seems likely that the complete explanation for the effects of actinin-3 loss will extend beyond actinin's structural role.[19]

In summary, we have explored three different ways that diverse actinin activities arise from extremely alike genes. Different alleles of the ACTN3 gene, though only divergent at one nucleotide position, produce proteins of dramatically different functional capacity. Paralogs of actinin, similar proteins coded for by related genes, reside in different locations and perform similar but not identical functions. Finally, each actinin paralog may execute multiple roles. Actinin certainly supports the sarcomere by binding actin, but it might also regulate muscle fiber type by interacting with calsarcin and other proteins.

What then of the ACTN3 genetic test? Does it provide useful information for athletes? Should parents test their children and use the results to nudge them toward one activity over another? Studies of elite athletes consistently show ACTN3-577X homozygotes to be underrepresented among sprinters and other power athletes. So if the Olympics are the goal, then the test might help eliminate certain events from consideration. But athletes should employ the ACTN3 test and other genetic tests with caution. Most studies show that only about 50 percent of exercise performance can be attributed to genetics—the rest depends on training, desire, diet, and other environmental factors.[20] And variation in ACTN3 explains only a small portion of the genetic effect. A recent study of non-elite Greek athletes showed that differences in ACTN3 explained no more than 2.3 percent of the variation in forty-meter sprint performance in males. If our genomes are like libraries, ACTN3 is only one volume.[21]

All in all, assessing ACTN3 alleles provides only a sliver of information about our capacities, so I cannot imagine discouraging a girl who loves to run from joining the cross-country team just because she has the wrong ACTN3 alleles. But as our understanding of genetics increases, tests will become more informative.

Even now, tests that assess panels of genes are replacing single-gene analyses. In Chapter 12, we'll examine the insights arising from the most recent work connecting genes to performance. The next chapter addresses myosin, actin's conspirator in power production.

Chapter 4

Not Too Fast, Not Too Slow

When I was in middle school, someone gave my family a bunch of old wooden baseball bats. Though the bats were in rotten condition, I loved them. These were serious adult bats, not aluminum lightweights that made tinny clinks. I sanded the least damaged one and applied a coat of mahogany stain. In the backyard I swung the restored bat, picturing myself at the plate in Veteran's Stadium, the Phillies' home at that time. I would imagine a sharp crack as the bat's graceful barrel met a hanging curveball, then the crowd's roar, and finally the sustained applause during my triumphant trip around the bases.

Things went differently when I took the bat down to the stickball field behind the cemetery. The bat was much too heavy for me, and I couldn't control it, so I rarely made good contact. Even worse, when I did get solid wood on the worn-out tennis balls we used, the results fell far short of my backyard fantasies. I couldn't generate enough bat speed to drive the ball, and my longest shots dropped short of the fence separating our ball field from the headstones. Before long, I ditched the heavy bat for a broom handle.

When I lucked into good contact with a skinny broomstick, the tennis ball sailed over the fence.

Hitting a home run depends on the ability to generate power. Indeed, the success of many athletic actions hinges on power, the combination of velocity and force. So athletes seeking power might believe they need to maximize *both* velocity and force. But when muscles contract at their highest velocities, force suffers. Conversely, when muscles contract at their highest forces, velocity wanes. To achieve peak power, athletes must find the perfect compromise between force and velocity. In a similar way, muscles have optimal lengths. At their shortest and longest, they produce little force. Muscles attain maximal force at intermediate lengths.

So each muscle has a sweet spot—an optimal length and velocity where power peaks. Athletes live for that sweet spot. We conjure it, nurture it, relish it. In the words of essayist John Jerome, finding the sweet spot is "so satisfying that it must be why we play."[1] In this chapter, we'll see that the key to the sweet spot resides in the properties of myosin, the molecular basis of muscular power.

In common usage, the word "contraction" implies shortening. But contracting muscles—ones actively producing force—can shorten, lengthen, or stay the same length. Pick up a dumbbell, let it hang with your arm extended, and begin a biceps curl. When you muster enough force to exceed the gravitational pull of the weight, your biceps shortens and the weight rises. Such contractions—those that occur while a muscle is shortening—are called concentric contractions. Now stop the curl with your elbow at a right angle, hold that position, and savor the effort.

Although your biceps is not changing length, you know from the burn that it is producing force. Such isometric contractions generate force with no change in muscle length. If you continue the isometric contraction long enough, your biceps starts to fatigue, and the weight begins to fall, stretching the biceps. The biceps now makes an eccentric contraction—one that occurs while a muscle lengthens. Completely relaxing the biceps when it starts fatiguing would be a mistake. The dumbbell would fall uncontrollably, possibly damaging your body or property. Maintaining an eccentric biceps contraction stabilizes and steadies the falling weight.

The biceps muscle inserts on the forearm near the elbow, far from the hand that holds the dumbbell. In other words, a lever arm—the forearm—intercedes between the place where your biceps applies force and the weight that it lifts. Compare the forearm to a crowbar. To lift a heavy rock, we can slip one end of a bar under the rock, place a pivot point near the rock, and then push down on the long lever on the other side of the pivot (Figure 4.1, above). In this example, we have two lever distances and two forces. The input lever (L_i) spans from the pivot to the end of the crowbar where we apply the input force (F_i); the output lever (L_o) spans from the pivot to the rock where the output force (F_o) emerges. The output force lifting the rock depends on the input force and the lengths of the two levers:

Equation 1: $F_o = \dfrac{F_i \times L_i}{L_o}$

Crowbars amplify force because the input lever is longer than the output lever. As a trade-off, big movements of the input lever produce only small movements of the output lever. So the crowbar enables us to hoist a heavy rock, but only move it a small distance. Because velocity is the distance traveled per unit time, the rock also moves slowly.[2] Equation 2 summarizes the relationship between velocity and lever lengths.

Equation 2: $V_o = \dfrac{V_i \times L_o}{L_i}$

Notice that the input-lever length switches position from the top of the fraction in the force equation to the bottom in the velocity equation, whereas the output lever flips in the other direction. These differences depict a trade-off between force and velocity. Changes that enhance force impair velocity, and vice versa.

Forearm levers differ from crowbar levers in two ways. One difference is that both input and output forces act on the same side of the pivot (Figure 4.1, below). Thus when the forearm is used as a lever, the biceps must exert an upward force to oppose the

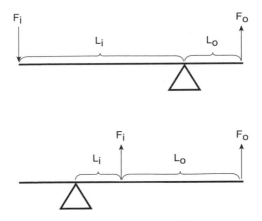

4.1. Lever diagrams. Above: Crowbar lever. Below: Human forearm lever.

downward gravitational pull on the dumbbell. More important, the biceps inserts on the forearm only a short distance from the elbow. As a result, the input lever is much shorter than the output lever—the distance between dumbbell and elbow. In other words, the ratio of input to output levers in the forearms is opposite the ratio in a crowbar lever. To lift a dumbbell, force produced by the biceps must substantially exceed the weight's gravitational force. But a shortening of the biceps translates into a large displacement of the dumbbell, so the dumbbell moves faster than the muscle. In short, human forearm anatomy favors velocity over force.

Although the forelimbs of all mammals have longer output levers than input levers, the ratio varies, reflecting trade-offs between force and velocity. Fast animals, such as horses, tend to have long forelimbs, resulting in long output levers, thereby maximizing velocity. In contrast, animals that dig, such as badgers, tend to have short forelimbs, resulting in short output levers. Also, the muscles of digging animals connect to bones farther from the joint, producing long input levers. Consequently, the forearm anatomy of digging animals emphasizes force at the expense of velocity.[3]

Though all humans display essentially the same shape, the ratio of our limb lengths to body mass does differ. Recent work shows that people with relatively longer legs run at higher efficiency, expending less energy at a given speed than short-legged folks.[4] Thus anatomical differences among humans influence their athletic talents. But no matter how much a runner desires horselike legs or a weight lifter seeks badgerlike arms, we can't alter our leg lengths or the locations where our muscles insert. On the other hand, we can control the position of our limbs, and the position also influences the amount of force we can produce because it alters muscle length.

To explore the effects of muscle length on force production, try this experiment. First, hold your left arm extended in front of your body, palm upward, and place your right palm on top. Now make a maximal isometric contraction of the left biceps, attempting to bend the left arm, but opposing it with pressure from the right. Next, repeat with the left arm bent at a right angle: pull your left elbow in toward your hip and extend your forearm in front of your body. Hold your left palm pointing upward, cover it with your right palm, and again contract the left biceps. Can you discern a difference in the amount of force produced by this contraction and the first one? Finally, repeat with your elbow fully bent. Point your elbow out away from the body like a chicken wing, and place your hand near chest, palm upward. Now press downward with the other palm.

The three positions in our example encompass the range of lengths traversed during a biceps curl: elongated at the start with the arm extended, intermediate as the elbow forms a right angle, and shortened with the weight fully raised and the elbow flexed. If you managed to contort yourself into these positions, you probably found that force production peaks at the middle length, dropping considerably when the elbow fully extends or bends. Though this demonstration was hardly a rigorous experiment, it mirrors the results of controlled studies on isolated muscles. The muscle force-length curve looks like an upside-down "U": intermediate lengths maximize force. So to attain peak force, athletes should take the Goldilocks approach. Muscle length should be "just right," rather than too long or too short. The narrow optimal range of muscle lengths points toward an advantage of vertebrate limb anatomy: our relatively short input levers mean that small changes in muscle length move limbs large distances, allowing muscles to remain close to their optimal length throughout contractions.

At the molecular level, interactions—called crossbridges—between actin and myosin proteins explain the force-length relationship. To see how these crossbridges work, we need to look at the structure of myosin. Myosin II—the dominant form of myosin in skeletal muscles—contains six separate protein strands: two myosin heavy chains and four myosin light chains. Heavy chains contain three distinct regions: a long thin tail, a small flexible neck, and a globular head (Figure 4.2).[5] Picture the tail as your forearm, the head as your bunched fist, and the neck as your wrist. Just as flexing and extending your wrist joint moves your fist, myosin necks can pivot, flipping the heads from one position to another. Two heavy chains, their tails entwined, form the core of each myosin II. Four myosin light chains, much smaller than heavy chains, hang onto the necks of the heavy chains. Although light chains regulate contractions in some tissues, they are less important in skeletal muscle, so we'll focus on the heavy chains.

Myosin heavy chains generate mechanical power through a crossbridge cycle composed of power and recovery strokes

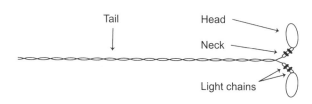

4.2. Myosin structure.

(Figure 4.3). Before the power stroke, myosin head groups fasten to actin. Their neck regions then swing like a lever arm and snap the heads into a new position. This motion tugs at the attached actin, producing force. Before the recovery stroke, myosin heads detach from actin. Head groups then return to their starting position, ready for another cycle. The detachment step is crucial to force generation. Imagine a rower who dips an oar in the water

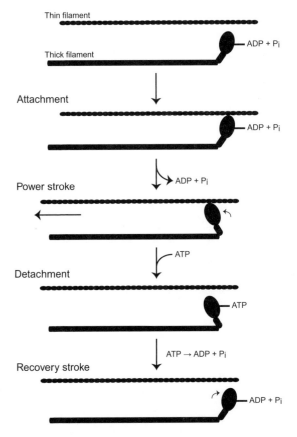

4.3. Steps in the crossbridge cycle. A single myosin heavy chain is depicted. Phosphate is abbreviated as P_i.

and pulls, then leaves the oar in the water and pushes back the other direction. The boat might oscillate a bit, but it won't make any progress. To be effective, the oar must leave the water on the return stroke. Similarly, myosin must attach to actin on the power stroke and then detach on the recovery.[6]

During the full crossbridge cycle, myosin heads bind actin, flip, detach, and recover, much as the oars of a rowboat dip into the water and pull, and then lift out of the water and return. And just as adding another rower increases the force pushing a rowboat forward, the amount of force generated by a muscle depends on the number of crossbridges. The ordered arrangement of skeletal muscle sarcomeres encourages crossbridge formation.

In sarcomeres, myosin-rich thick filaments interdigitate with actin-rich thin filaments (see Figure 3.1). At the center of each thick filament, myosins orient tail-to-tail, forming a central region devoid of head groups. About 300 regularly spaced head groups stud each thick filament, projecting like oars from an ancient rowing galley.[7] But if myosin heads are like oars, the thick filament is an odd boat. Its center lacks myosin heads, resembling a boat with oars only at each end. And oddly, the myosin heads at each end of the thick filament pull in opposite directions, like a boat with rowers on each end heaving in different directions. No muscular power would result if the two ends of a thick filament tugged against each other on the same thin filament. However, thick filaments don't futilely battle themselves. Rather, each end of the thick filament connects with a different thin filament, pulling the thin filaments toward each other, producing force that tends to shorten the sarcomere.

Just as forearm anatomy influences force production, the lengths of thick and thin filaments matter. Imagine thin filaments half as long as those in typical sarcomeres. Such stunted thin filaments would overlap less with thick filaments, resulting in fewer cross-bridges and less force. During the construction of sarcomeres, how do cells set thin filament lengths? One intriguing idea is that the giant protein nebulin may serve as ruler. To appreciate nebulin's size, let's compare it to other proteins. Each actin molecule weighs 42 kilodaltons (kDa, the most common measure of protein size).[8] Actinin weighs 100 kDa. Nebulin towers over these normal-sized proteins, tipping the scales at up to 900 kDa. Approximately 6,500 amino acids, encompassing 185 repeats of a 35-amino-acid sequence, link together to form each enormous nebulin. Remarkably, a single nebulin protein wraps around an entire thin filament, spanning almost 400 actin protein subunits. Because nebulin extends along the whole filament, it may act as a ruler, setting thin filament length.[9]

The lengths of thick and thin filaments don't change during exercise, but sarcomere lengths do (Figure 4.4). And when sarcomere lengths change, the number of crossbridges also changes, thereby affecting muscle force. At intermediate sarcomere sizes, thin filaments overlap the entire region of the thick filament that contains myosin heads. Thus in-between lengths result in the maximal number of crossbridges, optimizing force. When muscles stretch, the sarcomeres lengthen, and the overlap between thin and thick filaments declines. The number of crossbridges falls and so does force. At short muscle lengths, fewer crossbridges form for two reasons. Thin filaments overlap each other, blocking myosin heads from making attachments. Also, as thick and thin filaments push together, they deform, losing their regular parallel orientation. Force declines because fewer crossbridge attachments can form between these deformed filaments. At short lengths, crossbridge numbers may not be the only story—jamming filaments together can produce forces that oppose those generated by muscle contraction.[10]

Given the importance of sarcomere length, muscles must build sarcomeres of just the right size. Although lengthy nebulin may measure thin filaments, truly colossal titin, which weighs between 3,300 and 3,700 kDa, may measure sarcomeres. Each titin spans half the sarcomere, running from the Z lines at the sarcomere's boundary to the M lines at its center. During assembly of sarcomeres, titin establishes the distance between the Z and M lines and positions thick filaments at the sarcomere's center where they best produce force. Thus the length of each sarcomere is approximately twice the length of its titins. Titin also contributes to force production. When sarcomeres change length, the distance

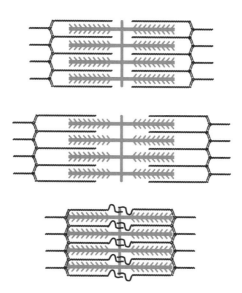

4.4. Effect of length on sarcomere function. Above: Sarcomere at optimal length. Middle: Lengthened sarcomere. Below: Shortened sarcomere.

between the M and Z lines changes, and titin stretches or compresses. Like a spring, titin pulls or pushes back, generating force in the opposite direction. Thus, because of titin's elasticity, stretched muscles produce some force without using energy. In this way, titin resembles the collagen of tendons, storing energy when stretched, then returning it as force.[11]

In summary, the parallel pattern of thick and thin filaments in skeletal muscle promotes force production by maximizing the contact between filaments and allowing many crossbridges to attach. But this regular pattern compels skeletal muscles to operate across a narrow range of lengths, because force drops precipitously outside a muscle's sweet spot. To achieve high force, skeletal muscle sacrifices versatility. Smooth muscles, in comparison, work well across a wide range of lengths. For example, after a Thanksgiving dinner indulgence, contractions of stomach smooth muscles propel foodstuffs out of the painfully stretched stomach, yet the same muscles also generate force after a small breakfast. Smooth muscle has more thin filaments than skeletal muscles and fewer—but longer—thick filaments. As a result, thick filaments overlap thin filaments even when the muscle stretches, making smooth muscle less sensitive than skeletal muscle to length changes. On the other hand, fewer thick filaments means fewer crossbridges and potentially lower force.[12]

Theoretically, athletic movements muster the most force when skeletal muscles operate near their optimal lengths. But do the muscles of athletes actually function at their ideal lengths? This question is tough to answer for at least three reasons. First, it is technically difficult to determine sarcomere length and force produced by muscles of exercising humans. Also, as we learned in Chapter 1, elastic tendons connect muscles to bones. The stretch and recoil of tendons complicates analysis of muscle length and force. Finally, most athletic movements employ a mixture of isometric, concentric, and eccentric contractions. When muscles contract through a range of lengths during an activity, they inevitably deviate from optimal length during at least part of the contraction. As an example, let's look at the movements of the foot during running and walking.

Contractions of opposing leg muscles power foot movements. The main calf muscle—the gastrocnemius—runs along the back of the lower leg. Recall from Chapter 1 that the gastrocnemius extends the foot, pushing it downward and away from the shin (see Figure 1.2). In contrast, the anterior tibialis runs along the shin and flexes the foot, drawing the foot upward toward the shin. During running and walking, these muscles activate alternately. The anterior tibialis contracts mostly during the swing phase, when the foot is off the ground, whereas the gastrocnemius contracts mostly during the stance phase, from the moment the foot hits the ground until push-off. Such alternating activation of opposing muscle pairs typifies cyclic activities such as swimming, running, walking, and cycling.

During running, the foot decelerates when it first hits the ground, and this braking creates forces that tend to flex the ankle. At the same time, the gastrocnemius contracts, producing force that tends to extend the ankle. Nonetheless, the gastrocnemius initially lengthens because the force tending to flex the ankle exceeds the force generated by the muscle in the opposite direction. In other words, the gastrocnemius undergoes an eccentric contraction at the start of the stance phase. This eccentric contraction stabilizes the ankle, controlling flexion and allowing the Achilles tendon to store elastic energy (see Chapter 1). During the later portion of the stance phase, the gastrocnemius shortens. This concentric contraction extends the foot and powers the push-off. So the gastrocnemius contracts through the entire stance phase, first making an eccentric contraction and then a concentric one.

Trail runners may gripe about uphills, but they fear downhills, especially late in long runs. Running downhill, the foot hits the ground with more force, so it takes stronger eccentric gastrocnemius contractions to decelerate the body and oppose ankle flexion. Similarly, stronger eccentric contractions of the quadriceps—the muscles on the front of the upper leg—are needed to oppose bending of the knee. Downhill running can be agonizing because eccentric contractions can damage muscle, leading to pain and soreness, especially when muscles stretch well beyond their normal lengths while contracting.[13]

Most muscles operate near—but not necessarily precisely at—their optimal lengths during exercise. To evaluate force-velocity relationships, scientists have employed a variety of methods, including measurements of joint positions, determinations of muscle activity, computer modeling, and imaging techniques that evaluate muscle length. A recent study of walking shows that the gastrocnemius starts near its optimal length, then stretches to slightly above the optimal length while the foot is on the ground, and finally shortens to considerably below the optimal length during push-off. Other muscles work slightly outside their optimal length throughout most of the walking stride. The vastus lateralis and medialis, both part of the quadriceps, operate at mostly shorter than optimal lengths. In contrast, the gluteus maximus, the largest muscle of the buttocks, remains mostly above its length optima.[14]

Running and walking use the same muscles but are quite different movements, so not surprisingly, studies find differences in muscle lengths between the two activities. Calf muscles are shorter during running than walking. Before foot strike during running, the gastrocnemius is below its optimal length. And though the gastrocnemius lengthens after foot strike, it probably remains below optimal length.[15] Although it seems inefficient for the gastrocnemius to work outside its optimal range, a somewhat shortened gastrocnemius may maintain tension on the Achilles, allowing it to store elastic energy during foot strike. During walking, elastic energy storage is not as important, and the gastrocnemius may therefore work closer to its optimal length.

Although force matters in athletics, power matters more in most events. Cyclists, swimmers, pole-vaulters, and placekickers perform very different actions, but they all seek to maximize power. Power drives the bicycle along the road, pulls the swimmer through the pool, launches the vaulter over the bar, and sends the football through the uprights. Across a broad range of sports, performance hinges on the capacity to produce power—the product of force and velocity:

Equation 3: *Power = Force × Velocity*

To contract powerfully, muscles apparently must apply high forces at high speeds.

Yet athletes don't always seek to move as quickly as possible. For an entire golf season, I silently chanted "slow, slow, slow" each time I stood over the ball. On the rare occasions when I managed to heed my own advice, my shots were not only more accurate, but also flew farther. Is it possible that slowing down improved my power? If so, does this effect tell us something about how muscles work? Certainly my struggles to find the right swing tempo represent a common issue in sports—pace is crucial across athletic events. But even though great players make it look simple, the golf swing is complex. So let's turn to a simpler example, cycling.

Pedaling rate, called cadence, strongly affects a cyclist's power output. Recreational cyclists often prefer cadences between 50 and 60 revolutions per minute (rpm). These cadences are too slow to optimize power, suggesting that recreational cyclists may rank something else, perhaps their comfort, above power. When scientists measure the relationship between cadence and power in a controlled laboratory setting, peak power occurs at cadences between 80 and 120 rpm, in line with cadences that racers adopt.[16] At lower cadences, cyclists can produce higher forces on the pedal, but the decreased pedal velocity leads to low power. When they attempt higher cadences, cyclists apply lower forces to the pedal, even when making a hard effort. For this reason, high cadences become especially difficult when pedaling against resistance (e.g., when going uphill). Like optimal muscle length, cycling cadence follows the Goldilocks principle: intermediate velocities produce the highest power. Bicycle gears allow cyclists to maintain an optimal cadence across a range of conditions.

Earlier in this chapter, we explored how our anatomy contributes to trade-offs between force and velocity. But the relationship between force and velocity also depends on the molecular interactions of actin and myosin. To see why, we need to examine the role of ATP in the crossbridge cycle.

ATP powers muscle contraction in two ways. First, at the end of the power stroke, ATP binds to myosin heads, causing them to

morph and detach from actin, thereby paving the way for the recovery stroke (Figure 4.3). Myosin detachment does not require
energy from ATP breakdown; ATP simply binds myosin, triggering its release from actin. This effect of ATP explains rigor
mortis—the stiffness that can last for a few days following death.
ATP stores run out a few hours after death. With no ATP available to promote detachment, crossbridges remain attached and
stiffness ensues.

After detaching from actin, myosins catalyze the breakdown
of ATP into ADP and phosphate, capturing some of the released
energy as movements of their necks. The slender myosin necks
swing like lever arms, flipping head groups into new positions,
completing the recovery stroke. Notice that myosins do not
burn chemical energy during power strokes to drive actin past
myosin. Instead, myosins split ATP during recovery strokes to
stretch myosin head groups into energized positions, a bit like
drawing back the strap of a slingshot. Power strokes generate mechanical force when myosin head groups flip back to a
relaxed position while bound to actin. So the power stroke resembles the release of a drawn slingshot, taking advantage of
previously stored energy. Thus the crossbridge cycle is another
example of energy conversions during exercise. Myosins temporarily store chemical energy as elastic potential energy of
myosin necks, then convert that potential energy into mechanical force.

An amazing technology—optical tweezers—allows scientists
to explore activity of individual myosin molecules. Scientists
attach both ends of an actin filament to small plastic beads and
hold the beads in place with focused laser beams (the tweezers).
They then expose the filament to a single myosin molecule, trig-

ger crossbridge cycles by adding ATP, and measure force and the distance traversed by actin, using the beads as markers. To show that movements of myosin's neck drive those of its head, scientists engineered myosin heavy chains with longer necks. Actin moves about 6 nm during a crossbridge cycle with normal myosin. Actin displacements increase proportionally as the neck region length increases, strongly supporting the neck as the hinge that drives the power stroke.[17]

Myosin cycles at about twenty times per second. If each crossbridge moves actin 6 nm, a single crossbridge can move actin no faster than 120 nm/sec. However, studies using intact thick and thin filaments show velocities as fast as 8,000 nm/sec. So the velocity produced by intact thick filaments greatly exceeds the speed produced by individual myosins. Attaining such high velocities requires that many myosins work together in intact filaments. Crucially, crossbridges cycle asynchronously on a thick filament, like a rowing galley where the rowers all follow a different rhythm. Although this would be chaos in a boat, it's an effective strategy in the thick filament. During the recovery stroke of one myosin, many others pull actin forward, so the actin filament moves much farther than the displacement of a single crossbridge. By rowing out of synchrony, the myosin heads of a thick filament generate higher velocities than those attainable by a single head. Plus, if myosin heads rowed synchronously, actin heads might slide backward during myosins' recovery strokes.[18]

Compare the activities of a thick filament to a tug-of-war. To generate the most force, a team should try to keep their hands on the rope and tug together. On the other hand, to move the rope quickly, team members should alternate their pulls, letting go of the rope while others tug. These scenarios resemble the force-velocity trade-off that myosin molecules face. To optimize force, myosin should remain attached to actin, maximizing the number of crossbridges. But to optimize velocity, individual myosins need to release actin so that their teammates on the thick filament can heave. Crossbridges lingering in the attached position resist the actions of other myosins, just as oars dallying in the water might oppose the pull of other oars. Thus duty ratio—the proportion of time myosin spends in the power stroke versus the recovery stroke—falls as contraction velocity increases because myosins detach quickly from actin, spending less time in the power stroke and more in recovery. Force falls at these low duty ratios, because fewer crossbridges are attached.[19]

Compared to myosins from other tissues, skeletal muscle myosin is a low duty-ratio motor, binding actin only 1–10 percent of the time. Earlier in the chapter, we noted that sarcomere structure helps skeletal muscles generate high forces. In this part of the chapter, we see that skeletal muscle's duty ratio favors velocity over force. Yet duty ratios of skeletal muscle myosin remain low even when velocity is low. For example, only about 10 percent of myosins connect to actin during isometric contractions. Why don't more crossbridges attach during isometric contractions, thereby producing higher forces? At any given moment, most myosin heads probably can't form crossbridges because they are not adjacent to a myosin-binding site, which occurs every 36 nm along the thin filament. In contrast, myosin heads bunch closer together on the thick filament, approximately 14.5 nm apart, and therefore may be distant from the nearest binding site on the thin filament. During a shortening contraction, myosin heads may need to pause at the end of the recovery stroke until a binding site slides close, making attachment possible. During isometric contractions, many myosin heads are probably unable to participate.[20]

Let's again compare skeletal muscle to smooth muscle, this time focusing on duty ratio. Smooth muscle generally makes slow, sustained contractions, such as the peristaltic contractions that move food from the stomach to the intestine. Although the low duty ratio of skeletal muscle myosins promotes velocity over force, smooth muscles contain myosin with high duty ratios. Thus smooth muscle myosins remain attached to actin through a larger portion of the crossbridge cycle, slowing velocity but enhancing force. Because of its high duty ratio, each smooth muscle crossbridge delivers more force than a skeletal muscle crossbridge, explaining how smooth muscle generates sufficient force despite having fewer crossbridges than skeletal muscle. Additionally, because fewer crossbridges are required to produce a given force, smooth muscle conserves energy, allowing sustained contractions.[21]

Our reductionist analysis in the second half of this chapter has focused on the properties of individual myosin molecules and sarcomeres. Yet each muscle cell contains millions of sarcomeres that together achieve much greater force and velocity than individual sarcomeres. Sarcomeres link in series, connecting end to end, and they run in parallel, lying alongside each other. The series arrangement of sarcomeres amplifies velocity. About 40,000 sarcomeres connect to span across a human muscle fiber. If each sarcomere shortens by 0.1 µm per second, a series of 40,000 sarcomeres shortens at 4,000 µm (about 0.15 inches) per second. The parallel arrangement of sarcomeres increases force. Although thousands of sarcomeres attach in series, billions run in parallel in large human muscles. Each muscle fiber contains approximately 5,000 parallel sarcomeres, and large muscles may contain a million or more fibers. Single sarcomeres produce only tiny forces, but muscles achieve high forces because the contributions of billions of parallel sarcomeres sum together.[22]

In the early and mid-1900s, many professional baseball players swung heavy bats. Babe Ruth's bats were supposedly as heavy as fifty-six ounces. Ted Kluszewski, playing for the Cincinnati Reds in the 1950s, reportedly hammered nails into his bats to make them heavier. More recently, players have taken steps to lighten their bats, hollowing the insides and stuffing them with cork or other fillers. Today's major leaguers typically swing bats lighter than thirty-four ounces. Analyses that combine the physics of bat-ball collisions with muscle physiology support this trend toward lighter bats.[23]

The distance a batted ball flies depends on many factors, but only a few can be influenced by the batter: swing speed, bat weight, and how well the bat contacts the ball. When a batter makes good contact, the ball flies farther both as bat weight and the bat head's speed increase. So sluggers might naturally turn to very heavy bats. But heavy bats require more force to swing, so swing speed suffers because of the force-velocity trade-off we've discussed. The best bat weight, like muscle length and shortening velocity, is somewhere between the extremes.

A mathematical model based on pitch speed and batter size predicts optimal bat weights ranging from thirty-seven to fifty-two ounces for professional players. Another model, one that incorporates the force-velocity trade-off of muscle, predicts lower optimal weights—between thirty-five and forty ounces. Although bat choice is a highly individual decision, most players choose bats lighter than these theoretical optima, probably because successful swings require more than power. Batters must decide whether to swing at a ball thrown at up to a hundred miles per hour (mph) from only sixty feet and six inches, and then connect the bat to the ball just as it crosses the plate. Swinging a lighter bat may sacrifice power, but bat speed increases, possibly giving the batter an extra fraction of a second before deciding whether to swing. Batters also have more control over lighter bats, so achieve solid contact more often when using them. So we can add yet another trade-off to our growing list: batters may sacrifice power to improve their odds of connecting solidly with the ball.[24]

Chapter 5

Lactic Acid Acquitted

Bliss transforms to anguish in an instant. One moment, I'm running hard, enjoying the speed, my body on autopilot. The next moment, pain suddenly erupts, not only in my legs, but in my lungs and arms too. Now I need sharp focus to maintain form and pace.

Every serious athlete has felt similar torment, and many blame lactic acid. According to widely held beliefs, toxic lactic acid floods the bloodstream during intense workouts and we feel the effects: a befuddled brain, shortness of breath, and burning agony in the limbs. Lactic acid, the usual story goes, also exerts wicked aftereffects. The day after a tough workout, lingering lactic acid produces tenderness, soreness, even pain. This notion of lactic acid as irredeemable villain is almost certainly false.

Controversy swirls around lactic acid—few subjects in exercise physiology generate as much debate. Intense arguments about lactic acid have recently appeared in the Point: Counterpoint series of the prestigious *Journal of Applied Physiology.* In one exchange, an author calls the reasoning of his antagonists "a continuation of a circular argument based on their own errors." Another group claims that the authors making the opposing case "choose to ignore more than a century of physical, chemical, and physiological science." These outbursts may be more polite than trash talk in a sandlot football game, but among professional scientists, they are fighting words.[1]

Words matter in scientific arguments. In fact, the quarrels about lactic acid often hinge on precise definition of terms such as "lactic acidosis," "anaerobic metabolism," and "lactate threshold." The phrase "lactic acid" itself generates controversy. Should we instead speak about the related compound lactate? In our body fluids, lactic acid falls apart into a negatively charged lactate molecule and a positively charged proton (H^+):

Equation 1: *Lactic acid $\leftrightarrow$ Lactate$^-$ + H^+*

The ratio of lactate to lactic acid in a solution depends on the solution's acidity—its concentration of H^+. In acidic solutions, Equation 1 tends toward the left, favoring lactic acid. In solutions with neutral and low acidity, lactic acid dissociates into lactate and H^+. Most body fluids have neutral acidities, so they contain much more lactate than lactic acid, a good reason to favor the term "lactate."

On the other hand, muscles make both H^+ and lactate during intense exercise, probably in similar amounts.[2] These compounds combine to form lactic acid in acidic solutions, justifying use of the term "lactic acid." Furthermore, by using the term "lactate" rather than "lactic acid," we neglect the H^+ ions produced by muscles. In the end, neither term perfectly captures the complex biochemistry of muscles during exercise. I use the shorter and friendlier "lactate" rather than "lactic acid" in the remainder of this chapter. Perhaps the caustic connotations of the phrase "lactic acid" contribute to the compound's overly negative reputation? And maybe increased use of the term "lactate" will point athletes and coaches toward a more evenhanded judgment of its role during exercise?

As discussed in Chapter 2, the phosphocreatine and ATP stored in our muscle cells powers only a few seconds of exercise. If you want to run a mile, your muscles will need to synthesize ATP using energy from nutrients: proteins, carbohydrates, and fats. You don't burn much protein until other fuel sources run low, for example, during starvation. Thus carbohydrates and fats—mainly glucose and fatty acids—propel exercise. This chapter focuses on glucose, a simple sugar containing six carbon, twelve hydrogen, and six oxygen atoms. Metabolic pathways convert high-energy glucose to lower energy molecules like lactate and carbon dioxide, using the released energy to attach phosphate to ADP, forming ATP. During a mile run, you extract ATP from glucose using two different systems: aerobic and anaerobic metabolism. We'll see later in the chapter that the phrase "anaerobic metabolism"—like lactic acid—has its shortcomings, but coaches, athletes, and scientists use it, so I will too.

Two distinct metabolic pathways conspire in anaerobic metabolism. Glycolysis, a chain of ten reactions, breaks one six-carbon glucose into two three-carbon pyruvates, yielding two ATPs (Figure 5.1). Then a single reaction—lactate fermentation—converts pyruvate to lactate, also a three-carbon molecule. Equation 2 summarizes the combined reactions of glycolysis and lactate fermentation:

Equation 2: *Glucose + 2 ADP + 2 Phosphate →*
$$2\ Lactate^- + 2\ H^+ + 2\ ATP$$

Neither glycolysis nor lactate fermentation requires oxygen; hence the phrase "anaerobic metabolism" to describe these reactions. Though this chapter focuses on lactate fermentation in human muscles, humans also exploit different types of fermentation

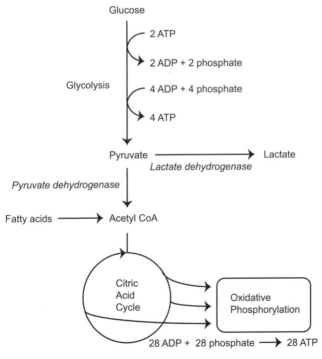

5.1. Overview of metabolic pathways.

performed by other organisms. For example, yogurt makers enlist the anaerobic bacterium *Lactobacillus* to ferment the milk sugar lactose to lactate. Beer makers recruit the yeast *Saccharomyces* to ferment sugars into buzz and bubbles—ethanol and carbon dioxide.

Aerobic metabolism, which requires oxygen, also begins with glycolysis. When oxygen-rich blood bathes muscles, pyruvate from the breakdown of glucose can flow into the citric acid cycle instead of lactate fermentation. The citric acid cycle completely picks apart pyruvate, transforming one six-carbon glucose to six one-carbon molecules of carbon dioxide and trapping energy in high-energy electron-carrier molecules. We'll see in Chapter 8 how a process called oxidative phosphorylation strips energy from these electron-carriers, yielding up to thirty-two ATPs.[3] Equation 3 describes the combined reaction for glycolysis, the citric acid cycle, and oxidative phosphorylation:

Equation 3: *Glucose* + 6 *Oxygen* + 32 *ADP* + 32 *Phosphate* →
6 *Carbon dioxide* + 6 *Water* + 32 *ATP*

Pyruvate sits at the crossroads of glycolysis, lactate fermentation, and the citric acid cycle, suggesting an analogy for metabolic pathways—they resemble roadways. Metabolic pathways intersect, merge, and branch. Some are lightly traveled back roads. Glycolysis, on the other hand, is a superhighway. Whether destined for the citric acid cycle or fermentation, glucose first flows into glycolysis, forming pyruvate. The junction of aerobic and anaerobic pathways at pyruvate resembles a traffic light at the center of town. Pyruvate can take a left and head toward lactate, or proceed straight into the citric acid cycle (Figure 5.1).

Suppose you could stand at that traffic light in your muscle cells and flip a remote, shunting pyruvate toward the citric acid cycle or lactate fermentation. Which direction should you send pyruvate during a mile run? Two factors play a role—time and intensity. Time influences your choice because anaerobic metabolism makes ATP more quickly than aerobic metabolism. After glycolysis, aerobic metabolism entails many additional steps: about ten reactions in the citric acid cycle, followed by a complex series of energy transfers in oxidative phosphorylation. In comparison, lactate fermentation requires only a single reaction. In the first ten seconds of exercise, your muscles must send plenty of pyruvate toward speedy lactate fermentation to ensure adequate ATP supplies while the slower gears of aerobic metabolism get cranking.

After the first ten seconds of exercise, the choice between aerobic and anaerobic metabolism hinges mainly on exercise intensity. If you plan a light one-mile jog, you should mostly favor efficient and clean aerobic metabolism. Aerobic metabolism yields about fifteen-fold more ATP per glucose molecule than lactate fermentation, so aerobic pathways use glucose sparingly, preserving fuel for later use. Also, aerobic metabolism transforms glucose to carbon dioxide, which the lungs readily remove from the blood, leaving no wastes behind. On the other hand, if you're brave enough to race the mile, you need to add some anaerobic metabolism to the mix. During high-intensity exercise, lactate fermentation contributes substantially to ATP production, providing an ATP turbo boost that helps you achieve your highest workload. Fortunately, our muscles decide on the ratio of aerobic and anaerobic metabolism without our conscious input—imagine needing to intentionally adjust metabolism during a soccer game or gymnastics routine.

If aerobic metabolism burns glucose cleanly and efficiently, does that make lactate fermentation dirty and inefficient? This view prevailed among scientists and athletes for much of the twentieth century. Many people pigeonholed lactate as a metabolic dead end, a harmful pollutant, a waste product formed when oxygen becomes limiting. According to this traditional story, athletes start to run low on oxygen as exercise intensity increases, so fermentation increases, leading to lactate accumulation and causing pain and fatigue. Though this account contains elements of truth, it unfairly emphasizes lactate's negative effects and misses some of its positive ones.

Runners racing the mile—or its modern metric substitute, 1,500 meters—produce lots of lactate. But though the 1,500 meter race offers lots of drama at track meets, scientists need more controlled exercise protocols to understand lactate's role. Graded exercise tests offer an orderly opportunity to investigate the body's response to different exercise intensities. A graded exercise test was one of my first activities upon joining an exercise physiology lab as a beginning graduate student. I had been playing some pickup hockey and I thought my fitness was respectable, so I welcomed the chance to show my new colleagues that I was prepared to contribute to exercise science not only academically but also physically.

My enthusiasm waned only a bit when I was ushered into a cramped environmental chamber, outfitted with a mask to measure my respiratory gases, and shown how to work the treadmill's emergency shutoff. After a light warm-up, my new colleagues increased the workload every two minutes. Although the initial intensities were easy, I soon was working hard. Before long, my legs and chest burned. Without a finish line to motivate me, I felt like quitting, but the encouragement from my colleagues—they were screaming at me—kept me going. Then the workload notched up again, and I knew I couldn't sustain the pace for a full two minutes. In addition to the physical pain, I felt the uncomfortable psychological sensation that the test precisely and unequivocally documented my limits. I wobbled off the treadmill with a healthy dose of empathy for the subjects who participate in exercise physiology studies.

Scientists make numerous different measurements during graded exercise tests. We'll focus on two: blood lactate and oxygen consumption (VO_2).[4] To measure VO_2, scientists outfit subjects with a mask, like the one I wore during my graded exercise test. The mask captures expired air and sends it to instruments that measure oxygen concentration and airflow. To calculate VO_2, scientists subtract the oxygen levels in expired air from those in the inhaled air and multiply by the volume of air passing through the lungs. VO_2 corresponds to the rate of aerobic metabolism because it assesses the amount of oxygen the body consumes while producing ATP.

When we plotted the results of my graded exercise test, VO_2 rose progressively as exercise intensity increased and then abruptly plateaued (Figure 5.2). Your results would look similar. The oxygen consumption plateau corresponds to the maximum rate of oxygen consumption (VO_{2max}), a measure of an athlete's total capacity to make ATP through aerobic pathways. Athletes with strong aerobic fitness attain high VO_{2max} values; they also reach VO_{2max} at higher workloads than less fit athletes. Once an athlete reaches

VO_{2max}, any additional boost in workload must result from lactate fermentation.

To measure blood lactate during a graded exercise test, scientists sample blood with a pinprick and then assess its lactate concentration. Inexpensive, portable lactate meters make lactate testing a widely used method. Blood lactate follows a very different pattern than oxygen consumption, remaining low at the lowest workloads, rising only slightly at moderate workloads, and then increasing considerably at the highest workloads. The exercise level corresponding to the first appreciable increase in blood lactate is the lactate threshold. Most people reach lactate threshold at workloads between 60 and 80 percent of the intensity corresponding to their VO_{2max}. Scientists and coaches sometimes use the phrase "anaerobic threshold" rather than "lactate threshold." However, blood lactate levels do not exactly track anaerobic metabolism, so I stick with the phrase "lactate threshold."

Although good science demands the precision of blood lactate measurement, you don't need a blood test to know when you

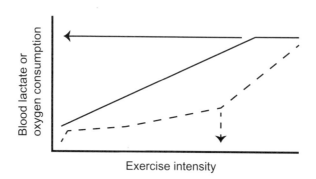

5.2. Typical results from graded exercise test. Solid line represents oxygen consumption; dotted line represents blood lactate level. Arrows indicate VO_{2max} and lactate threshold.

breach the lactate threshold. As you cross it, your focus narrows, your breathing accelerates, and your perception of exertion sharpens. In short, exercising above the lactate threshold hurts. These symptoms confirm that the lactate threshold marks a physiological transition.

Earlier, I compared metabolic pathways to roads. Another helpful analogy compares a metabolic pathway with a factory assembly line. Some metabolic pathways construct molecules like the assembly lines that build cars. But other metabolic pathways, including the ones that we discuss in this chapter, break down glucose, or dismantle molecules. So picture a *dis*assembly line, with glucose taken apart step by step. Each station along an assembly line receives materials, modifies them, and passes them along to the next station. The output of one step becomes the input for the next step. Likewise, the product of each enzyme along a metabolic pathway is the reactant for the next enzyme. In this way, metabolic pathways link reactions together, each enzyme providing the starting material for the next enzyme in the pathway.

Let's consider just two linked reactions. If all is well, the second enzyme rapidly uses the first enzyme's product. This keeps the concentration of the first reaction's product low and thereby drives the first reaction forward. (Recall from Chapter 2 that concentration of reactants and products strongly influence a reaction's direction.) If the second reaction stops, however, product from the first reaction accumulates, opposing the first reaction. Such effects can ripple through entire pathways. Inhibit the last enzyme in a pathway and the products of all of the previous reactions build up, so the entire pathway grinds to a halt.

Can we use this simple model of metabolic pathways to explain the lactate threshold? Picture yourself performing a graded exercise test. At low and moderate workloads, aerobic metabolism fulfills your muscles' ATP demand. Pyruvate, the product of glycolysis, flows happily into the citric acid cycle, fueling oxidative phosphorylation, which consumes oxygen while it makes ATP. Now, as your exercise intensity increases, suppose that your blood no longer delivers oxygen as fast as your muscles use it. Muscle oxygen levels fall and pain intensifies. You have reached VO_{2max}. At this point, sending more pyruvate toward the citric acid cycle causes a metabolic traffic jam—the buildup of intermediate products of the aerobic pathway.

After attaining VO_{2max}, you may not want to work any harder, but you probably can. Because it does not require oxygen, you could run glycolysis faster, leading to more ATP synthesis and pyruvate production. But if the extra pyruvate can't enter the citric acid cycle, it could accumulate, clogging glycolysis. So lactate fermentation becomes an attractive detour, removing pyruvate and saving glycolysis from gridlock. In this scheme, blood lactate production rises when oxygen falls. This framework supports the traditional view of lactate as waste product, produced only when aerobic metabolism has reached its capacity.

An apparent problem with this interpretation is that most athletes cross their lactate threshold before reaching VO_{2max}. In other words, lactate begins to accumulate before aerobic metabolism reaches its maximum. This accumulation might be explained if oxygen levels vary across muscle cells. Perhaps the first oxygen-limited muscle fibers produce lactate before other fibers reach their aerobic limit. A more serious problem with the notion of lactate as waste product is that this interpretation rests on an erroneous assumption—that blood lactate level solely depends on the rate of lactate production. Actually, the lactate threshold arises from the interaction of two processes: lactate production and consumption.

Consider this analogy: At a fundraising pancake breakfast, a cook slides stacks of flapjacks under a heat lamp. Volunteers grab the pancakes and serve hungry customers. Just as both lactate production and consumption determine blood lactate level, the number of pancakes under the lamp depends on how quickly the cook flips pancakes and how voraciously the diners devour them. Growing piles of pancakes might be the result of either an especially productive cook or a sparse crowd in the dining room.

Is there a metabolic equivalent of a room full of hungry diners, removing lactate from blood? To answer this question, scientists employ techniques that estimate lactate production or consumption rates. Measuring blood lactate won't suffice because blood lactate level won't change if its production and consumption both increase to the same extent. But scientists can measure lactate consumption by injecting chemically tagged lactate into the blood and watching its disappearance. Such studies show that lactate fermentation can reverse when oxygen is available, converting lactate back to pyruvate. Pyruvate then enters the citric acid cycle, which dismantles it to carbon dioxide, eventually yielding lots of ATP. Scientists have known for a long time that lactate fermentation runs backward after intense exercise, clearing lactate from the blood by burning it aerobically.

More recently, physiologists discovered that a "lactate shuttle" runs during exercise: exercising muscles produce lactate and the blood carries this lactate to the heart, brain, and even other skeletal muscles. These other tissues then consume lactate through aerobic pathways. When we exercise, contracting muscles make more lactate than they do at rest, but other tissues compensate by consuming lactate more quickly. During light exercise, blood lactate increases only slightly because the extra consumption keeps up with the additional production. During high-intensity exercise, lactate production increases considerably because aerobic metabolism can no longer keep up with the muscle's demand for ATP. We reach lactate threshold at the exercise intensity where lactate production greatly outstrips its consumption.[5]

Thus our muscles produce lactate not just as a last-ditch effort to make some additional ATP when our oxygen supply stretches to its limit. Rather, some muscles make lactate even at low exercise levels and then ship it over to neighboring tissues that burn it. At least for some cells, lactate is a fuel, not a metabolic dead end. And the lactate shuttle is not a sideshow—huge amounts of lactate flow through it. During exercise, lactate probably provides most of the energy for certain tissues. These findings provoke in me some sympathy for lactate, which is so often demonized, even as it works hard to feed our cells during exercise.[6]

Wondering why we produce lactate in one place and then shuttle it to another? The lactate shuttle definitely does not supply *extra* fuel to our bodies because all the lactate we burn ultimately comes from glucose breakdown. Instead, the shuttle connects fermentation in one tissue to aerobic metabolism in another, probably with benefits for both tissues. Tissues with low aerobic capacity—such as fast-twitch muscle—run glycolysis followed by lactate fermentation, producing ATP and lactate. Rather than accumulating, this lactate passes along to tissues with high aerobic capacity—such as heart, brain, and slow-twitch muscle—which burn it as a convenient fuel. One tissue's trash becomes another tissue's treasure.

Some cancer cells, like the cells of some exercising muscles, synthesize lactate when oxygen is plentiful. One explanation for this similarity is that both cell types use ATP rapidly. While muscle cells need energy to contract, cancer cells require ample energy for their rapid growth and division. Another explanation is that both cell types sometimes face anaerobic conditions—tumors often lack adequate blood supply. Cells that run lactate fermentation because they sometimes experience anaerobic conditions may also tend to make lactate in aerobic conditions. Thus making lactate in the presence of oxygen may be a side effect of making it in oxygen's absence. Whatever the explanation, scientists are exploiting this metabolic difference between cancerous and normal cells for both diagnosis and therapy.[7]

In summary, lactate threshold does not correspond to the onset of lactate fermentation, and lactate fermentation does not occur only when oxygen levels fall. In fact, muscles often robustly produce lactate in the presence of oxygen, especially during low- and moderate-intensity exercise. Thus the label anaerobic metabolism to describe glycolysis followed by lactate fermentation is misleading—another instance of treacherous terminology. Widespread use of the phrase "anaerobic metabolism" may have hindered athletes, and even some scientists, from appreciating how much fermentation happens when oxygen is abundant.

Most tissues—for example, livers and kidneys—burn ATP at a fairly constant rate. In contrast, muscle ATP consumption increases up to a hundred-fold during exercise. If the liver is like an adult cat lazing away a sunny Saturday, then muscle is a kitten—apparently unconscious one moment and then somersaulting hysterically the next. Despite the dramatic swings in ATP usage, muscle ATP levels remain quite steady.

Had enough graded exercise tests? Let's run a hundred-meter dash instead. As we stand in the starting blocks, our muscle ATP concentration is about 5.2 millimolar (mM). Crossing the finish line, ATP has dropped to about 3.7 mM. The small fall in ATP might contribute to the fatigue we experienced at the finish. On the other hand, the rapid use of ATP threatened to drop ATP levels much farther. Muscle's capacity to preserve relatively stable ATP levels in the face of dramatically different activity levels is an example of homeostasis—maintenance of constant conditions in the face of changing demand or environment. To achieve homeostasis, muscle cells must regulate the metabolic pathways producing ATP—accelerating ATP synthesis at the start of exercise and throttling it back during rest. Both aerobic and anaerobic pathways start with glycolysis, so the best way to control ATP levels is to regulate glycolysis. Just before exercise, glycolysis saunters. Seconds later, it sprints.[8]

ATP levels powerfully influence glycolysis through a simple negative feedback mechanism: high-ATP concentrations shut down glycolysis, preventing even more ATP from being made. Conversely, when ATP decreases, glycolysis activates, restoring ATP levels. ATP influences glycolysis by controlling enzymes in the glycolytic pathway. But in linked pathways like glycolysis, not every enzyme needs to be regulated, just as cities don't need to install traffic lights every ten yards. Instead, cells control only key enzymes, particularly those at the start of pathways and at branch points.[9] For example, the enzyme phosphofructokinase catalyzes the first *committed* reaction in glycolysis. Before the phosphofructokinase reaction, intermediate products can branch off to other pathways. Afterward, they are on a turnpike without exit ramps and must flow through glycolysis. By modifying phosphofructokinase activity, cells can control the rate of glycolysis.[10]

ATP controls phosphofructokinase by binding to the enzyme and contorting it, a process called allosteric regulation. When ATP sticks to a regulatory site on phosphofructokinase, the negative charges on ATP influence the shape of the enzyme, morphing it into an inactive form.[11] ATP molecules tend to bind phosphofructokinase when ATP concentration rises, but fall away when ATP level falls. In this way, phosphofructokinase responds to changing levels of ATP. ADP levels have the opposite effect, activating phosphofructokinase by attaching to regulatory sites and contorting the enzyme into an active form.[12] As we begin running the hundred-meter dash, myosin starts consuming ATP, so ATP levels fall and ADP levels rise, activating phosphofructokinase. Hence, glycolysis and ATP synthesis increase.

Let's move now to the end of glycolysis and examine the multienzyme complex called pyruvate dehydrogenase. Pyruvate dehydrogenase catalyzes the conversion of three-carbon pyruvate to two-carbon acetate, which can enter the citric acid cycle.[13] So think of pyruvate dehydrogenase as the traffic light at the ramp from glycolysis to the citric acid cycle. If it's green, pyruvate flows toward aerobic metabolism. If it's red, pyruvate enters lactate fermentation.[14] Like phosphofructokinase, pyruvate dehydrogenase activity depends on its shape, but the mechanisms that control pyruvate dehydrogenase differ from the allosteric regulation of phosphofructokinase.

Two enzymes determine whether the pyruvate dehydrogenase traffic light glows red or green. A kinase catalyzes the attachment of a phosphate group to pyruvate dehydrogenase. Phosphate groups carry negative charges, so their attachment contorts pyruvate dehydrogenase, flipping the traffic light to red. Conversely, a phosphatase enzyme flips the light to green, activating pyruvate dehydrogenase by removing the phosphate (Figure 5.3).[15]

The modification of enzymes by opposing actions of kinases and phosphatases is a type of covalent regulation because kinases attach phosphates to their targets with sturdy covalent bonds. When a kinase adds a phosphate to pyruvate dehydrogenase, making the stoplight red, the enzyme remains dormant until a phosphatase removes the attached phosphate, flipping the light to green. Hormones often modulate enzymes by covalent modification. For example, before the hundred-meter dash, while we stand nervously in the blocks, hormones may use covalent regulation to turn on enzymes that activate ATP synthesis. Our

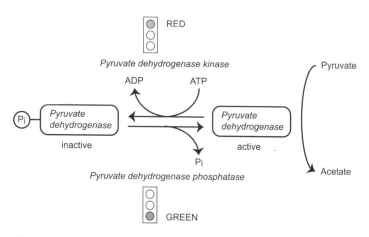

5.3. Regulation of pyruvate dehydrogenase. Phosphate is abbreviated as P_i.

muscles therefore begin to make extra ATP even before its level falls. In contrast, allosteric activation of phosphofructokinase does not occur until exercise begins and ATP and ADP levels change.[16]

Regulation of pyruvate dehydrogenase offers an alternative to oxygen availability as an explanation for the pattern of lactate production during exercise. As exercise intensity increases, glycolysis activates, producing more pyruvate. Pyruvate dehydrogenase converts much of this pyruvate to acetate, which then flows into the citric acid cycle. But if pyruvate dehydrogenase does not activate enough to transform all the pyruvate into acetate, then the remaining pyruvate shunts toward lactate fermentation. At moderate exercise intensity, some lactate fermentation probably occurs because glycolysis rate somewhat exceeds pyruvate dehydrogenase activity. This production of lactate occurs due to the regulation of pyruvate dehydrogenase, rather than the absence of oxygen. At higher exercise intensities, glycolysis may produce much more pyruvate than pyruvate dehydrogenase can convert to acetate, leading to rapid lactate production. In this mechanism, the number and activity of pyruvate dehydrogenase enzymes, rather than the presence or absence of oxygen, determines lactate production.[17]

Lactate itself may play a role in regulating metabolism by serving as a signal that intense exercise is under way.[18] For example, lactate may trigger dilation of blood vessels, stimulating blood flow to exercising muscles. Lactate also stimulates collagen synthesis, possibly contributing to remodeling of tissues in response to exercise. Finally, lactate may oppose the breakdown of fats during intense exercise, causing muscles to favor carbohydrate rather than fats as a fuel source.[19] In Chapter 9, we discuss the balance of fat and carbohydrate metabolism in more detail.

In summary, muscles produce lactate long before oxygen becomes limiting, other tissues use this lactate as a fuel, and lactate may be a signaling molecule. But lactate's roles as fuel and signal do not necessarily exonerate it as a source of pain and fatigue. Lactate might be cellular nourishment at low levels, but poison at higher ones, in the same way that low levels of caffeine enhance exercise performance, whereas high levels can be deadly. Put simply, the dose might make the poison.

Some evidence points toward lactate as a source of fatigue. Lactate threshold correlates with endurance. A classic study measured the endurance of cyclists with high- and low-lactate thresholds. Scientists measured thresholds as a percentage of each athlete's own aerobic capacity (VO_{2max}). Those with high thresholds (~80% of VO_{2max}) accumulated only half as much blood lactate and pedaled twice as long as those with low thresholds (~65% of VO_{2max}). Furthermore, many studies show that endurance training shifts the lactate threshold toward a higher workload, resulting in lower blood-lactate levels during exercise. This work unquestionably connects lactate to fatigue. The tempting conclusion—and one of the traditional claims about lactate—is that high blood lactate *causes* fatigue.[20]

Yet to blame fatigue on lactate based on this evidence is to confuse correlation with causation. Water hoses, sirens, spectators, and cigarettes are often found in proximity to house fires, but not all are causes of fires. Hoses and sirens are tools for battling fires, not their cause. We've already seen lactate in the role of siren, as a signal that triggers responses. Maybe lactate is also like a hose, part of the machinery that offsets fatigue, rather than fatigue's cause? Or perhaps lactate is like a spectator, neither a cause of fatigue nor a part of the system for fighting it. On the other hand, just as cigarettes are a cause of fires, lactate could be a cause of fatigue. In summary, lactate could cause, signal, offset, or merely be correlated with fatigue. We need rigorous experimental tests to differentiate among these possibilities.

Unfortunately, assessing the effects of lactate in humans poses difficult experimental challenges. In particular, scientists struggle to untangle lactate's effects from other physiological perturbations during exercise. One useful strategy is to artificially elevate blood lactate by infusing lactate. Athletes who receive lactate infusions feel no adverse effects during moderate exercise even when their lactate levels reach more than double the normal levels, arguing against lactate as the cause of fatigue. Most animal experiments also indicate that lactate is benign; some suggest it may be beneficial. Bathing isolated muscle in lactate can even lead to enhanced contraction force. Though controversy remains, these studies suggest that lactate may be more like a water hose than a cigarette, countering fatigue rather than causing it.[21]

Perhaps protons, rather than lactate, cause fatigue? Proton concentration alters protein structure, so increased acidity during exercise could impair function. Experiments testing this hypothesis paint a complex picture. Rising acidity weakens performance of muscles in some experiments, but this effect typically vanishes when physiologists perform the experiments at normal body temperature. Further, increased acidity seems to be protective against other factors that harm muscle performance. High acidity, for example, can offset the negative effects of increased potassium ions on muscle contraction.[22]

So what causes fatigue? Changes in salt concentrations—potassium, calcium, or phosphate—might disrupt muscle function. When the going gets tough, muscles also release chemicals called metabolites that could cause pain and fatigue. We may also fatigue because energy supplies run low. Or our brains might be to blame: neural and psychological impairments definitely accompany fatigue. In the end, it might be quixotic to seek a single explanation for fatigue. Multiple factors probably contribute, and the sources of fatigue may vary across exercise intensities and durations.[23]

Recent science exonerates lactate, casting it as an earnest contributor rather than a defiant scoundrel. Scientists no longer see lactate as a poisonous pollutant, but instead as a fuel, protectant, and signal. Although lactate's rags-to-riches story might mirror a good made-for-TV movie script, the renovation of lactate's reputation threatens to overturn decades of training advice based on lactate. Do we need to completely rethink training? Fortunately, no.

In *Blink,* Malcolm Gladwell shows that we often make good decisions instinctively and later rationalize them with apparently logical but completely invented explanations.[24] Perhaps something similar is at work when coaches justify long runs as a way to flush lactate out of tired muscles and prevent muscle soreness. This rationale does not hang together. Lactate quickly clears from muscles during exercise and recovery, so no lingering lactate remains to be flushed away twenty-four hours after exercise. And we've seen that lactate accumulation is not necessarily harmful. Nonetheless, the long run is sound training practice. Slow distance runs increase aerobic capacity and shift the lactate threshold toward higher intensities, improving endurance. Plus, long runs—

especially with friends, teammates, or even a good audiobook—nourish the soul.

Though blood lactate does not reliably match lactate production, it does offer a glimpse into an athlete's metabolism. Performance of elite endurance athletes correlates highly with lactate threshold, so training intended to increase the threshold makes sense. And lactate threshold can be used in simple ways to guide training: long runs below it, mid-length tempo runs at it, and short interval workouts well above it. But taking an overly simplistic view of lactate should be resisted. Despite more than a hundred years of inquiry into lactate during exercise, controversy and complexity persist. For example, a precise definition of lactate threshold remains contentious—a final instance of semantic slipperiness in the lactate debates.[25] Moreover, individual differences among athletes complicate simple applications of lactate threshold to training. Some folks tolerate elevated lactate; others crumple when lactate rises.

In the end, the complex physiology of lactate during exercise does not invalidate it as a training tool. Our expanding understanding of exercise metabolism might even improve its use, so stay tuned for the next developments.

Chapter 6

Catch an Edge

On a sunny January morning, I watched my sons duck between trees, catch air, and carve graceful turns. Snowboarding looked easy. I began to wonder whether I should try it. My skiing experience must count for something, I rationalized. Still, I pledged to start slowly—no inverted aerials on the half-pipe until I got comfortable. And though I didn't expect to need it, I grabbed a helmet from the rental hut to set a good example for the boys.

What I really needed was thick tailbone padding. Novice snow-boarders must lean back, weight the board's uphill edge, and skid cautiously. Staying off the downhill edge is the cardinal rule—offenders are slammed down like victims of a nasty judo throw. Afraid of catastrophic wipeouts, I stubbornly overemphasized the uphill edge, repeatedly lost my balance, and inevitably dropped onto my backside. I went home comprehensively bruised.

By my third day of boarding, satisfying signs of progress emerged. My turns became smoother and my speed increased. On the last run, with my sons watching from the bottom, I was in fine form. For a moment, I even started enjoying myself rather than obsessing about the uphill edge. The next instant, my down-hill edge gripped the snow, hurling me groundward. My body hit first, whipping my head into an icy patch. The helmet prevented serious physical injury, but the psychological damage was done. My snowboarding career was over.

Any number of lessons might be drawn from my snowboard-ing misadventure, but in this instance my point is about the com-plexity of athletic movements. Even seemingly simple athletic acts demand tremendous dexterity. If you've tried snowboarding, I'm sure you'll agree. If you haven't, recall your first shanked five iron, missed free throw, or strikeout. Learning a new sport can be humbling—and even painful.

The doctor taps your knee with a hammer, triggering a stretch reflex. Sensors in your knee detect the hammer's strike. A message—sudden stretch!—flows upward from your leg to your spinal cord. The spinal cord translates this message into a plan of action: contract the quadriceps and relax the hamstring. Direc-tions flow down to the leg, and you kick. This knee-jerk response illustrates the steps of a reflex arc: sense, decide, and respond. Three types of neurons (nerve cells) embody these steps. Sensory neurons detect stimuli and carry signals toward the central ner-vous system—the brain and spinal cord. Interneurons connect one neuron to another, forming circuits that make decisions. Finally, motor neurons carry instructions to target organs such as muscles and trigger responses.

A simple circuit of neurons underpins the stretch reflex (Figure 6.1). Sensory neurons relay signals from stretch receptors in the quadriceps to the spinal cord. These neurons directly stimulate motor neurons that carry signals back to the quadriceps, activating them. Thus abruptly stretching the quadriceps triggers their contraction through the simplest circuit possible: signals pass directly from the sensory neuron to the motor neuron. The same sensory neurons also activate interneurons that inhibit the hamstring's motor neurons, thereby preventing contractions that would op-

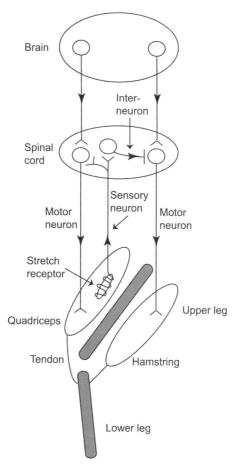

6.1. Muscles and nerves involved in the stretch reflex. Arrows indicate the direction of action potential propagation. V-shaped neuron endings indicate excitatory synapses; flat lines indicate inhibitory synapses. Bones are shaded.

pose the quadriceps contraction. The interneuron converts a stretch signal from the quadriceps into inhibition of the hamstring. Because of these spinal cord connections, the quadriceps contract powerfully in response to stretch while the hamstring relaxes, resulting in a quick kick. This involuntary reflex occurs at the level of the spinal cord, with no involvement from the brain.[1]

Uncoordinated, unconscious knee jerks appear to have little in common with the graceful, intentional moves of experienced athletes. Yet complex voluntary movements, such as snowboard aerials, follow the same steps as simple involuntary ones. (Yes, snowboarding jumps are voluntary, though peer pressure may sometimes make them feel compulsory.) To execute a switch-frontside 1080 double cork, you first sense, then decide, and finally respond: you see the ramp, hear your board scraping across hard snow, and sense the position of your limbs. These sensory data flow to the complex circuitry of your brain, which integrates the incoming information, compares it to your intention, and decides upon a course of action. Many motor neurons then transmit instructions to activate dozens of muscles and fine-tune their contraction strengths. Thus we can describe the snowboard aerial using the same sense-decide-respond framework as the knee jerk. However, landing a snowboard jump requires more complexity in each step. And unlike knee jerks, control of muscles during a snowboard jump isn't a one-and-done event. The control pathways run nonstop, continuously adjusting muscle contractions as you approach the ramp, pop the jump, spin gracefully, and nail the landing.

Voluntary movements and involuntary reflexes not only share similar processes, they also share some of the same neural circuitry. When it directs voluntary movement, the brain stimulates the same motor neurons triggered by the stretch reflex. Thus signals flowing down to the spinal cord from the brain inter-mingle with signals flowing upward from the muscles. Informa-tion from the brain can modify the stretch reflex, and the stretch reflex can modify instructions from the brain.

One consequence of the interactions between instructions from the brain and simple reflexes: stunningly fast movements that nonetheless retain some precision. When we were growing up, my brother and I played street hockey, trading the goalie and shooter positions, often firing point-blank slapshots at each other (and sometimes, admittedly, at our brave younger sister). Though I enjoyed drilling the ball into a corner and watching the net rip-ple, I especially relished the extraordinary feeling that accompa-nied the occasional great save—the sensation that my body moved automatically, even before I consciously noticed the ball's trajec-tory. While instructions from the brain initiate a kick save, the brain need not laboriously micromanage every detail. Rather, in-structions from the brain cooperate with extremely fast reflex pathways, enabling the goalie's leg to kick almost instantaneously.

The stretch-reflex circuitry also coordinates slower move-ments. For example, because activation of one muscle also trig-gers inhibition of muscles with antagonistic actions, the circuitry encourages alternating contractions of opposing muscle pairs such as the quadriceps and hamstring. This pattern underpins cyclical motions like running, cycling, and swimming. Additionally, sig-nals about muscle length flow up to the brain, contributing to our awareness of joint position. Finally, the circuitry helps fine-tune muscle contractions, reconciling idealized instructions from the brain to physical realities. For example, we will see that the stretch reflex helps snowboarders and trail runners respond to subtle ter-rain variations.[2]

Data travel long distances in both voluntary and involuntary pathways. In the stretch reflex, for example, signals travel between the knee and the spinal cord. During snowboard jumps, instructions flow from the brain to leg muscles. In both cases, information moves as electrical signals—called action potentials—along nerves.

A typical neuron contains a cell body, bushy processes called dendrites, and a long thin projection called an axon (Figure 6.2). Dendrites detect sensory stimuli and receive messages from other neurons. Cell bodies integrate incoming signals. Axons carry messages, sometimes across great lengths. Thus each neuron operates like a miniature reflex arc, receiving information, making decisions, and sending signals.

The longest axons transmit signals about a meter, from the base of the spinal cord to muscles in the foot. Superficially, axons

6.2. Stylized representation of a typical neuron. Not drawn to scale.

resemble your home's electrical wires—both are skinny cables that carry electricity—but their mechanisms differ. Electric current flows to your lamp and computer when negatively charged electrons move along metal wires. These wires conduct electricity effectively because their insulation prevents current loss on the way to your lamp. In contrast, current declines quickly as it travels along uninsulated axons. Without a boost, messages from the spinal cord would deteriorate long before reaching the foot. To counter signal decay, axons carry information in the form of action potentials that fire repeatedly as they pass down axons, allowing signals to traverse long distances. During action potentials, ions—charged molecules—flow across cell membranes through membrane proteins. Hence, action potentials carry signals by allowing ions to pass out of cablelike axons, whereas electrical wires work by preventing electron leak.

Cell membranes—like the walls of a house—are barriers, separating interior from exterior, preventing the promiscuous movement of materials. Membranes partition inside from out, because they are composed of phospholipids—molecules with a fatty, water-fearing tail and a charged, water-loving head. Two layers of phospholipids form cell membranes (Figure 6.3). Their heads point toward the watery milieu either inside or outside the cell, while their tails avoid water by pointing inwards toward each other. This lipid core blocks movements of molecules, especially charged ions, that avoid fatty environments.

Membrane proteins—like doors, window screens, and plumbing—regulate the transport of substances, making it possible for internal conditions to differ from external ones. Two classes of membrane proteins convey ions across axon membranes: pumps and channels. Channels produce pores in the membrane, creating paths for ions to flow passively. Pumps actively ferry ions, generating differences in ion levels across cell membranes and establishing the forces that drive ions through channels.

Sodium pumps carry three sodium ions out of the cell in exchange for two potassium ions, accumulating sodium outside cells and potassium inside (Figure 6.4). Sodium pumps also create membrane voltage—potential energy resulting from differences in charge across the membrane—because pumps push three positive charges out of the cell but only pull in two. In this way, the sodium pump makes the inside of cell membranes slightly negative compared to the outside. To power ion transport, sodium pumps split ATP, releasing chemical energy. Thus sodium pumps convert chemical energy into differences in ion concentration and charge across the membrane. These ion imbalances are potential energy, analogous to gravitational energy. Imagine the sodium pump pushing sodium up a hill outside the cell and pulling potassium up a hill inside the cell.

Electrochemical gradient—the combined influence of electrical and concentration imbalances—determines how ions flow through channels. When sodium channels open, sodium tumbles

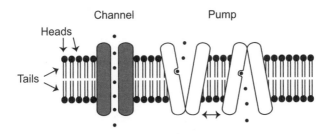

6.3. Diagram of a channel (shaded) and pump (open). Small dots represent ions. Two configurations of the pump are shown, one facing each side of the membrane.

down a steep path *into* cells because both concentration and electrical gradients favor sodium entry. In contrast, when potassium channels open, potassium rolls down a flat path *out of* cells through potassium channels because a strong concentration gradient favors potassium exit from cells, but a somewhat weaker electrical gradient opposes its exit.[3] At the start of action potentials, sodium channels open, allowing positively charged sodium ions to flow into the cell, thereby increasing membrane voltage. Then, sodium channels close and potassium channels open. The resultant potassium efflux returns membrane potential toward its starting level.

Picture yourself as a National Hockey League goalie. You strap on twenty pounds of padding and take your place between the pipes. A skater crosses the blue line, fires a slapshot, and a hard rubber disk flies toward you at 100 mph. Now would be a bad time to start generating electrochemical gradients. Pumps work slowly; so long before you established the basis for an action

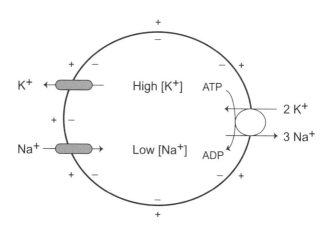

6.4. Channels, pumps, and ion concentrations in a typical neuron. Shaded cylinders represent channels; open circle represents pump.

potential, the shooter's stick would be raised in celebration. In contrast, ions pass quickly through channels. Thus pumps slowly generate gradients; channels rapidly exploit them. For this reason, neurons constantly run their sodium pumps, maintaining gradients of sodium and potassium and thereby remaining ready to fire action potentials. The functional differences between channels and pumps stem from their dissimilar structures.

Portions of pumps and channels—transmembrane regions—embed deep in cell membranes. Transmembrane regions resemble barrels bobbing in a lake, connected to each other by lengths of rope. Imagine some of the ropes dangling into the lake, others connecting barrels above the water's surface. In this analogy, the lake surface resembles the cell membrane, the barrels represent transmembrane regions, and the ropes represent intracellular and extracellular loops. Transmembrane barrels reside comfortably in the membrane's fatty interior because they contain water-fearing hydrophobic amino acids. Intracellular and extracellular loops possess charged amino acids that exist most happily in the watery environment inside and outside the cell.

Although pumps and channels both contain transmembrane regions, their structures and mechanisms of action differ substantially. To picture the difference, imagine a box with a divider down the center and fifty ants and fifty ladybugs in each compartment. The divider represents the cell membrane and the insects represent ions. To sequester all the ants on the left and all the ladybugs on the right, we would need to grab animals, lift them across the barrier, and then release them. Likewise, pumps bind ions on one side of the membrane, morph so that the ions face the other side, and then release the ions. This bind-flip-release mechanism, like carrying individual insects across a divider, occurs slowly and requires energy. Two distinct structural domains enable each sodium pump to transport ions: an intracellular domain that captures energy by catalyzing ATP breakdown, and a membrane domain that binds and releases ions.

With your palms facing each other, touch your wrists together and pull the tips of your fingers apart, forming a "V." Your hands represent the sodium-pump membrane domain; your arms represent the cell membrane. Ion-binding sites reside on your palms. Now, touch your fingers together and pull your wrists apart, causing the binding sites on your palms to point toward the other side of the membrane. This maneuver illustrates a vital property of the sodium pump—small shape changes flip its ion-binding sites between inside and outside the cell. The intracellular domain powers these contortions with the energy released when it splits ATP.[4]

Sodium attaches enthusiastically when its binding sites on the pump face inside the membrane, but it falls off and diffuses away from the cell when the sites face out. So as the binding sites flip across membrane, the pump attaches to sodium inside the cell and releases it outside, thereby carrying sodium out of the cell. On the other hand, potassium readily binds to outside-facing sites and detaches when its binding sites face inside, so the pump moves potassium into cells. The sodium pump resembles trains that allow only men to ride into the city, while permitting only women to ride out. When such trains run, women soon become overrepresented outside the city, just as sodium pumps accumulate sodium ions outside the cell. Potassium ions, like the men riding the trains, would follow the opposite pattern, accumulating inside the cell.

Let's quickly snap out of the molecular world and return to the ice rink—remember, an incoming puck screams toward you at

100 mph. Whether you choose to save the slapshot or to get the heck out of the way, you have less than a second to integrate sensory information, decide how to react, and activate your muscles.[5] Fortunately, action potentials travel even faster than slapshots— over 250 mph in some neurons. Action potentials propagate speedily for two reasons: (1) Action potentials themselves occur quickly. During action potentials, flow of ions through channels shoots membrane voltage from negative to positive then back to negative in about one millisecond. (2) Action potentials swiftly trigger new action potentials in nearby patches of axon membrane. So a train of action potentials sprints down your axons, carrying your brain's commands fast enough to orchestrate a spectacular save. Channels underpin the rapid initiation and transmission of action potentials.

To envision how channels work, let's again consider our box of insects. Suppose that all the ladybugs are in the right compartment and all the ants are in the left. If you punch a hole in the wall dividing the compartments, ants will move from left to right in their random wanderings. (Assume that ants and ladybugs ignore each other.) As ants begin to gather on the right side, some will occasionally move back to the left. But the overall movement will be from left to right until approximately equal numbers of ants reside on each side. Overall movement of ladybugs will be in the other direction until their numbers also equalize between left and right.

Similarly, channels provide pathways for ions to diffuse along existing gradients, powered by the ions' own random movements. On the other hand, our simple hole-in-a-wall analogy fails to capture two key characteristics of channels. First, channels transport specific molecules. For example, sodium ions flow easily through sodium channels, but other ions do not readily pass through them. Also, channels can be regulated—sodium channels open and close in response to tension, voltage, and other signals. To update the box analogy to reflect these properties, envision that the hole dividing the left and right compartments allows ants to pass, but not ladybugs, and comes equipped with a gate that can shut, sometimes preventing even ants from passing.

Sodium channels display remarkable selectivity. They preferentially carry sodium over potassium ions, even though the ions have similar size and identical charge. Ions with two positive charges, such as calcium, do not move through sodium channels at all. The basic unit of a sodium channel contains six transmembrane barrels. A complete channel contains four repeats of this six-barrel unit, for a total of twenty-four transmembrane barrels (Figure 6.5).[6] Sodium ions flow through a pore formed by eight barrels, two from each unit. Picture these barrels encircling a water-filled cylinder, holding the surrounding membrane at bay. Selectivity resides in loops connecting the pore-forming barrels. These loops stick into the pore, narrowing it into a selectivity filter. Mutations to specific amino acids in this filter abolish the sodium selectivity, allowing both calcium and potassium ions to pass through the channel.

Selectivity is one reason why membrane potential changes rapidly during action potentials. If sodium channels allowed potassium ions to pass, potassium flowing out of the cell would blunt the effects of sodium moving in, thus slowing the action potential. Another reason is that ions flow about a thousand-fold more quickly through channels compared to pumps, much as we might expect ants to pour through an opening faster than we would be able to pick them up and move them over a barrier. Under ideal conditions, approximately ten million sodium ions can pass through a sodium channel each second. Finally, many sodium channels activate at once during the action potential. To understand this last point, we need to see how sodium channels are regulated.[7]

The sodium channels that generate action potentials react to changes in membrane voltage, opening when voltage becomes more positive. The voltage sensor of sodium channels resides in the sixteen transmembrane barrels that do not contribute to the pore. In particular, buildup of positive charge inside the membrane

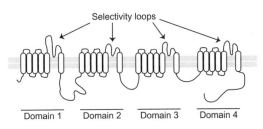

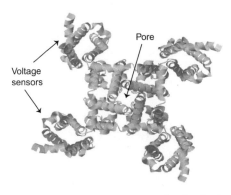

6.5. Structure of sodium channels. See note 6 for details about this structure.

repels the positively charged amino acids in four barrels, one from each unit. As a result, these barrels slide upward, shifting away from the cell interior and twisting the channel in a way that opens the pore. In this way, buildup of positive charge inside neurons opens voltage-sensitive sodium channels during action potentials.[8]

Action potentials contrast with the homeostasis—constant internal conditions—that cells normally maintain. Although negative feedback preserves homeostasis, positive feedback drives fluctuations of membrane potential during action potentials. Positive feedback happens when the response to a stimulus enhances the stimulus, triggering progressively bigger responses. I've seen it in youth soccer matches: a team scores a goal, gains some confidence, and starts playing better, leading to another goal that further increases confidence. Likewise, action potentials reflect positive feedback regulation of sodium channels. Action potentials begin when a stimulus makes the membrane voltage slightly more positive. This change opens voltage-sensitive sodium channels and allows sodium ions to flow into the cell, causing the membrane voltage to become even more positive and thereby instigating the opening of yet more sodium channels. Opening a few sodium channels thus triggers many more sodium channels to open. As a result, the neuron membrane quickly becomes highly permeable to sodium, which flows into the cell, causing membrane voltage to soar (Figure 6.6).

What goes up, of course, must come down. If voltage remained high, no further action potentials could fire. Two processes drive voltage back to its starting level. Sodium channels automatically close shortly after opening, halting flow of sodium into the cell. At about the same time, voltage-sensitive potassium channels open, providing a path for potassium to flow out of the cell. This loss of positive charge restores membrane voltage back toward its negative resting level. Often, potassium channels remain open for a few milliseconds, creating an "undershoot" that temporarily decreases the likelihood of another action potential.

Action potentials are all-or-nothing events. Although their shape varies somewhat across cell types, each cell fires a distinctive action potential that varies little in magnitude or duration. In this way, action potentials resemble smoke signals, ones where every puff is identical.[9] Both communicate with a one-letter alphabet; only frequency can be varied. Also, both are effective over limited distances. Because a smoke signal in New York is not detectable in Los Angeles, a whole series of fire pits would be required for a coast-to-coast smoke message. Similarly, although the effects of an action potential can spread to nearby membrane, these

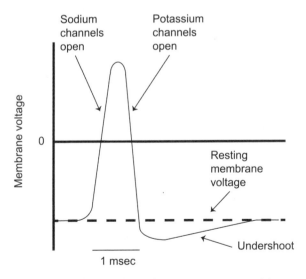

6.6. Membrane voltage during an action potential.

effects decay quickly. To send commands from the brain to the calf, action potentials must fire progressively along an axon.

During action potentials, some of the sodium ions that enter through voltage-sensitive channels diffuse along the axon, increasing voltage of nearby membrane. This rising voltage causes voltage-sensitive sodium channels to open, stimulating a new action potential adjacent to the first one. The process repeats, each action potential triggering another one next to it, and a train of action potentials passes down an axon. Although this system allows signals to propagate without decay, it would be very slow if every bit of membrane needed to fire an action potential, just as smoke signals would take forever to cross the country if fire pits were stationed a yard apart. A speedier system would position fire pits at the maximal distance where puffs from other pits could be reliably detected. Even better would be to provide binoculars to the operator of each station to increase the distance between pits.

In axons, a fatty substance called myelin acts like binoculars, extending the distance between action potentials. Myelin insulates axon membranes, preventing ions from flowing across them, except at occasional nodes with exposed membrane. Current flows quickly and with little decay along the insulated sections of axon in the same way that electrical current moves along an insulated wire. So an action potential at one node rapidly causes an increased membrane voltage at the next node, triggering an action potential there. Action potentials thus skip along an axon like smoke signals between well-spaced fire pits. As a result, action potentials carry signals across the body quickly and without degradation.

Action potentials are the messengers of the stretch reflex, carrying signals to and from the brain. What sets these messengers into action? And how do they eventually provoke muscles into contracting?

Recruit a friend to help you with this demonstration of the stretch reflex: hold your arm out, palm up, with your elbow at a right angle. Now, ask your assistant to drop objects of different weights—a Wiffle ball, an orange, perhaps a lead ingot if you have one lying around—into your hand. As the object hits, try to keep your hand from moving. Occasionally, you will judge the object's weight perfectly, and your hand will remain still. More often, your hand will deflect slightly upward or downward before quickly correcting back to its original position. The deflections will be bigger if you close your eyes or if your partner drops objects whose weights are difficult to predict, such as the Styrofoam "rocks" that were tossed around on the set of the original *Star Trek* series. Practice, especially with the same object, will reduce deflections.

Two distinct processes operate in this simple experiment. First, as the object drops toward your hand, you make a judgment about its weight and messages race from your brain to your biceps, ordering a contraction. This response is predictive—your brain makes an educated guess about the needed contraction strength. Then, after the object hits your hand, feedback pathways adjust the strength of contraction, correcting for errors in the original prediction. Biceps stretch, for example, triggers stronger biceps contractions, restoring the muscle to its starting length—the stretch reflex in action. Also, stretch sends sensory signals racing to the brain, where they influence output to the biceps and other muscles. For example, the brain might recruit other arm muscles to help stabilize your hand. If the weight is heavy enough, muscles of the torso and legs might even be asked to help.

Though the above experiment is contrived, it demonstrates that feedback from muscles modifies instructions from the brain. Such modification is needed when external conditions are different from expected: when you step in a hole during a trail run, when your opponent is just a bit stronger than you thought, and when the wind suddenly gusts during your backswing. Feedback from muscles must also modulate signals from the brain when your body changes, for example, when your muscles become weaker because of fatigue or stronger because of training.[10]

How do our muscles detect stretch? Skeletal muscles contain two types of cells: force-producing fibers and stretch-detecting fibers. Force-producing fibers—the workhorse cells—move bones. Stretch-detecting fibers collaborate with sensory neurons to translate stretch into action potentials. Sensory neurons that wrap around stretch-detecting fibers contain specialized sodium channels.[11] Stretch pulls these channels open, allowing sodium ions to flow in, increasing membrane voltage. In response, voltage-sensitive sodium channels open, triggering action potentials that flow up to the spinal cord and brain. So two types of sodium channels in the sensory neurons collaborate to generate signals: stretch-sensitive channels open in response to tension, producing small increases in membrane voltage that trigger action potentials by opening voltage-sensitive channels.

During snowboarding, the quadriceps shorten and lengthen. If stretch-detecting fibers were passive, they would feel great tension when the muscle lengthens, eliciting sensory signals similar to those during the knee-jerk reflex. But we don't jerk our knees with every turn of the snowboard because stretch-detecting fibers contract and relax along with force-producing fibers during voluntary movement. So when quadriceps contractions go as expected, both types of fibers undergo similar changes in length. As a result, tension on stretch receptors remains unchanged and sensory signals stay constant. When a snowboarder expectedly hits uneven terrain, signals from the brain fail to keep stretch-detecting fibers at a constant length. Tension on stretch receptors changes, opening or closing stretch-sensitive channels, generating signals in the sensory neurons. These sensory signals flow to the spinal cord, where they modify instructions from the brain, thereby adjusting contractions to account for unpredictable events.[12]

When movements fail to meet expectations, contraction strength must be fine-tuned. The response may be to push harder against a stronger-than-expected opponent or go lightly across a patch of softer-than-expected snow. How do we adjust force? Try this simple demonstration: hold your palms together in front of your chest, pressing them lightly together. Slowly increase the pressure by contracting your arm and chest muscles. Multiple motor neurons innervate these muscles, with each neuron activating a portion of one muscle's force-producing fibers. Recruiting additional motor units—motor neurons and all of the muscle cells they activate—produces progressively stronger contractions. During a maximal contraction, all of a muscle's motor units activate.

Do the next experiment outside, or perhaps only in your imagination: hold an egg with your fingertips and make a few vertical leaps without dropping or crushing it. Then place the egg between your wrists and repeat. Try again with the egg between your elbows. OK—jumping with your elbows pressed together is awkward. But the point is that when you jump, the egg accelerates, and just the right force needs to be applied to avoid crushing it. We control finger contractions more precisely than those of the chest and arm, so hold the egg between your fingers to avoid a mess.

Finger muscles are more precise than chest muscles partly because their motor units are small. Each motor neuron activates only a few finger muscle cells, allowing the number of activated cells to be fine-tuned. When a single motor unit fires, only a few cells contract, resulting in a weak contraction. Activating additional motor units produces small stepwise gradations in contractile strength, making delicate tasks like writing—and egg-holding—possible. Motor units of the arm and chest muscles are much larger, with each motor neuron innervating hundreds or even thousands of muscle cells; thus control of their contractile strength is coarser.

Do fingers also regulate contractile strength by modulating the strength of individual muscle contractions? To evaluate this mechanism, we need to appreciate how muscles activate. Let's consider a simplified system. Suppose we purify thin and thick filaments (see Chapter 3), mix them in a test tube, and observe. Nothing happens. They probably need some energy, so let's add some ATP. Still nothing. But when we add some calcium ions, crossbridge cycling ensues. In the absence of calcium, a thin protein called tropomyosin lies across actin, blocking crossbridges from forming. The protein troponin clamps tropomyosin onto actin. When troponin binds calcium, it morphs in a way that pulls tropomyosin away from actin's myosin-binding site, allowing crossbridges to form. Calcium unlocks muscle contraction, so let's see how muscles regulate their calcium levels.

In relaxed muscles, the sarcoplasmic reticulum—membrane-bound sacs spread throughout muscle cells—sequesters cellular calcium (Figure 6.7). Calcium pumps, structurally similar to sodium pumps, pack the sarcoplasmic reticulum membrane. These pumps expend ATP to move calcium into the sarcoplasmic reticulum. As a result, calcium in the sarcomere lowers, preventing crossbridges and thereby promoting muscle relaxation. Also, calcium in the sarcoplasmic reticulum rises, preparing the muscle cell for a contraction. When channels in the sarcoplasmic reticulum open, calcium flows out, rapidly raising calcium levels surrounding the contractile proteins and thereby triggering a contraction. Note the similarity to axons: the calcium pump of muscle, like the sodium pump of neurons, sets the stage for a rapid response by creating gradients that channels exploit.

Calcium triggers muscle contractions, but what causes the sarcoplasmic reticulum to release calcium? Cell membranes of muscles resemble those of axons—they contain voltage-sensitive channels. Action potentials in motor neurons stimulate action potentials that flow along muscle membranes, signaling calcium channels in the sarcoplasmic reticulum to open. Calcium rapidly flows down

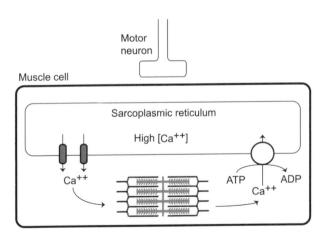

6.7. Activation of muscle contraction. Shaded cylinders represent calcium channels; open circle represents calcium pump.

its gradient, binds to troponin, crossbridges form, and the muscle contracts. Calcium pumps, always at work, pump calcium back into the sarcoplasmic reticulum, and the muscle relaxes.

So action potentials travel down motor neuron axons, stimulating muscle action potentials that open calcium channels of the sarcoplasmic reticulum. The resultant pulse in muscle calcium triggers a twitch—a rapid contraction followed by relaxation. Twitches, like action potentials, are invariant events—one twitch looks about the same as another. On the other hand, motor units modulate twitch rate. During exercise, motor neurons usually fire action potentials at high frequency, so new twitches begin before previous ones have ended, fusing jerky twitches into smooth contractions. During high-frequency stimulation, summation occurs—fused twitches produce more force than single ones. Hence, force increases progressively with increasing frequency of motor-neuron action potentials. In summary, we fine-tune muscle contractions through two mechanisms: (1) precise recruitment of motor units and (2) variations in the frequency of action potentials in the recruited motor neurons.

Movements of ions across membranes not only regulate contraction, but also factor into fatigue. The establishment of electrochemical gradients by sodium and calcium pumps demands substantial amounts of ATP during exercise and recovery. The myosin heavy chain, which splits ATP to drive myosin-actin crossbridge cycling (Chapter 4), consumes about 60–70 percent of the ATP used during a typical contraction. The other 30–40 percent goes toward ion movements that regulate contractions, mostly to fuel the calcium pumps that sequester calcium in the sarcoplasmic reticulum. Muscle sodium pumps expend a small amount of ATP to establish the sodium and potassium gradients that enable muscle action potentials. Under some circumstances, the energy required to regulate contractions may exceed that used to produce force. For example, in muscles that contract and relax very quickly, such as rattlesnake tail muscle, calcium pumping may consume ATP fiftyfold faster than it does in slower muscles. Even in ordinary muscles, the ATP required to control contractions can be a significant drain on energy resources.[13]

Because potassium flows out of cells during action potentials, its level outside muscle cells can rise from 4 millimolar (mM) up to 13 mM because action potentials increase during exercise. Conversely, sodium accumulates inside cells. These disturbances alter membrane voltage, reduce electrochemical gradients, and inactivate channels—all effects that can contribute to fatigue by impairing muscle activation. Sodium pumps typically work extra hard during exercise, increasing their activity up to twenty-fold to offset ion imbalances. On the other hand, very high intensity exercise inactivates sodium pumps. This shutdown could be an attempt to funnel all available energy to force generation, though the trade-off is rapid development of fatigue-causing ion disturbances. The endurance of isolated muscle preparations declines markedly when scientists inhibit the sodium pump, emphasizing the importance of pumps in maintaining muscle function. Compared to couch potatoes, athletes have extra muscle sodium pumps, so athletes more effectively counter the rise of extracellular potassium and intracellular sodium.[14]

In this chapter, we connected molecules to motion across many levels of biological organization, from proteins to people. At the protein level, pumps and channels usher ions across membranes through transformations in their shapes. At the cellular level, activity of these proteins provokes action potentials and calcium fluctuations. At the tissue level, cells cooperate in stretch receptors to detect tension, in neural circuits to make decisions, and in motor units to activate specific muscle cells. At the organ level, whole muscles—calves, quadriceps, hamstrings, and many others—contract. And finally, at the whole person level, coordinated muscle contractions produce both knee jerks and snowboard tricks.

Although making these connections is satisfying, the story I've told is anything but complete. Many other proteins, including other ion transporters, contribute. Another type of stretch receptor, the Golgi tendon organ, plays a role. The stretch reflex is only one of several responses mediated by circuits in the spinal cord. And, most obviously, I've only touched upon the role of the brain—the topic of the next chapter—in coordinating output to motor neurons.

Chapter 7

Your Brain on Exercise

Some people cry at the movies. I cry at finish lines. More than once, worried race volunteers have approached me. "Doing fine, these are tears of joy," I say, and they scurry off to find someone whose problems are physical rather than mental. Though I've learned to deal with these crying bouts, they still embarrass me. I'm well aware that finishing fifth in my age group at a local half marathon doesn't justify tears of any sort. But a hard thirteen-mile run definitely tweaks my brain chemistry.

I don't need a race to mess with my mind; a simple trail run will do the trick. My body and the environment—both in the background when I'm sitting at my desk—assert themselves on the trail. My calves scream on uphills, my quads squeal on the way down. My feet probe beneath fallen leaves. Hole? Rock? Snake? My body propels my brain down the trail, continually renewing its sensory input: first, the sweet song of a thrush; then, the sharp scent of pine needles; and later, a cool breeze rustling the emerald leaves of an oak.

On the trail, my attention sometimes flows out from my brain into my moving body, even projecting into the woods to unite with the oaks and thrushes. For a moment, I'm part of an ecosystem, one living thing connected to many others. Other times, my mind completely escapes the physical world, and I run miles without noticing my legs or surroundings. Instead, I compose a lecture, design an experiment, or outline the next chapter, sailing past obstacles that seemed insurmountable in my office. And the effects of exercise continue after I've showered and returned to my desk. Occasionally, a pleasant runner's high lingers. Often, my focus and memory are enhanced. Almost always, I'm more cheerful. In short, trail running alters my consciousness.

Recent research affirms my experiences. In fact, science increasingly portrays exercise as a wonder drug: producing nice feelings, improving memory, and protecting our brains against the ravages of stress, aging, and disease. Exercise, according to the emerging story, makes us smarter and happier while it buffs our bodies. And though exercise activates some of the same pathways as drugs like marijuana and heroin, it sidesteps (or even counters) some of their nasty effects. Exercise affects the brain by modifying synapses—the structures that connect neurons to each other and their targets.

Human brains contain about 100 billion neurons, each linking to thousands of other neurons, all packed into a space about the size of a grapefruit.[1] The dazzling complexity of neural connections enables shortstops to lay down bunts, snag line drives, and decide whether to throw to first or home after fielding a grounder. But it presents a daunting challenge for neuroscientists who seek to understand how our brains coordinate movement. Consider a shortstop reacting at the crack of the bat to make a diving catch. Somehow, the shortstop's brain tracks the ball's path, decides where the body needs to move, programs a sequence of muscle contractions, and updates these instructions as the ball flies. How can we unravel the neural pathways that accomplish these tasks? A useful approach has been to identify functional units: to discover brain regions specialized for particular tasks and to map the circuits between them.

Picture the brain as a multilayer cake. The bottom layers handle basic jobs, whereas the top layers perform the most sophisticated duties. At the base is the brainstem, responsible for controlling breathing, digestion, and heart contractions. In the middle are several regions, including the cerebellum, which helps coordinate movement. At the top is the cerebrum, filling more than three-fourths of the human brain and responsible for thinking, planning, memory, and other complex tasks.[2] The uppermost part of the cerebrum—the brain's icing—is the thin, deeply folded cerebral cortex.

Cerebral cortex—made of sensory, associative, and motor regions—initiates voluntary movement. Sensory cortex collects sensory input and sends it to association cortex, areas that assimilate the data and hatch plans. The shortstop's association cortex integrates auditory information (the crack of bat on ball) and visual information (the path of ball leaving bat) and makes a decision (dive). The premotor cortex translates this objective into a program of muscle contractions and informs the motor cortex. Finally, the shortstop dives when motor-cortex neurons send signals to the spinal cord, activating motor neurons that innervate muscles. So, in the simplest case, a linear pathway activates voluntary movements: sensory cortex to association cortex to premotor cortex to motor cortex to spinal cord to muscle (Figure 7.1).[3]

But relying on such straightforward circuitry would probably lead to more errors than outs, so it is no surprise that quality-control systems modify the brain's instructions to muscles. The cerebellum and basal ganglia—which are part of the cerebrum—both adjust signals as they pass from association cortex to premotor cortex, fine-tuning the planned sequence of muscle contractions. The cerebellum also modifies signals after they leave the motor cortex, reconciling output with sensory feedback from muscles. The brainstem has a role in motor coordination too, accepting input from the cortex and cerebellum and sending axons

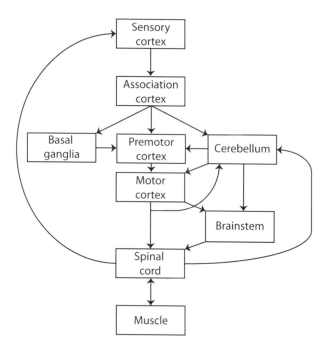

7.1. Neural pathways in voluntary movement.

down to the spinal cord. Finally, as we learned in Chapter 6, spinal circuits process instructions. In fact, most projections from the brain connect to spinal-cord interneurons rather than directly activating motor neurons. Synapses form the connections between neurons in all these pathways.

Two neurons—a sending cell and a receiving cell—kiss at synapses. These synaptic kisses are unlike passionate smooches between consenting adults. Instead, they resemble cheek pecks: the affection flows one way from kisser to recipient. At each synapse, sending neurons kiss receiving neurons, but receiving neurons usually don't reciprocate. On the other hand, most neurons receive synaptic kisses at some synapses and send them at others. Neurons typically pass information in one direction—they collect signals at synapses on their cell body and dendrites, and they distribute output at synapses at their axon endings.

Another difference between synaptic kisses and passionate smooches is that synapses send chemical, rather than physical, messages. Sending cells secrete neurotransmitters that diffuse across a thin gap and bind to receptors on receiving cells. A further difference is that synapses are stable structures. Sending cells persistently hover above receiving cells like overeager suitors, always ready to pass along chemical kisses. About 100,000 synapses speckle the cell body and dendrites of a typical motor neuron (see Figure 6.2). So, if we push the kissing analogy, many thousand friends, relatives, and admirers surround each motor neuron. When kisses from these acquaintances sufficiently excite it, the motor neuron passes the love along its axon to the hundreds of muscle cells it innervates.[4]

Most synapses convert electrical signals in a sending cell to chemical signals in the gap between cells and then back to an electrical signal in the receiving cell. Neuromuscular junctions—the synapses between motor neurons and muscles—follow this pattern. Action potentials (electrical signals) in the motor neuron trigger the release of the neurotransmitter acetylcholine (chemical signals), resulting in muscle-cell action potentials (electrical signals) that eventually stimulate muscle contractions.

Motor neuron endings puff chemical kisses toward muscle cells. At each neuromuscular junction, the motor neuron's axon widens into one or more buttonlike swellings crammed with small membrane-bound spheres, called vesicles. Each vesicle contains about 3,000 molecules of acetylcholine (Figure 7.2). When action potentials flow down the motor neuron axon, they trigger the opening of voltage-sensitive calcium channels in these axonal swellings, causing calcium to rush in. As in muscle cells (Chapter 6), calcium acts as chemical messenger: it binds to proteins and morphs them, activating machinery that fuses approximately a

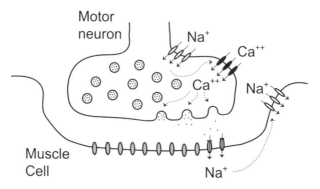

7.2. The neuromuscular junction. Gray cylinders represent acetylcholine receptors; open cylinders represent voltage-sensitive sodium channels; black cylinders represent voltage-sensitive calcium channels; small dots represent acetylcholine molecules. Dotted lines on right side of the figure depict steps in the activation of the synapse.

hundred vesicles with the axon membrane. As a result, the motor neuron releases about 300,000 acetylcholine molecules toward the muscle cell on the other side of the synapse.[5]

Muscle cells stand poised to receive acetylcholine kisses from motor neurons. Across from motor-neuron endings, acetylcholine receptors pack muscle-cell membranes. Like all receptors, acetylcholine receptors convey signals across cell membranes, translating extracellular signals into intracellular effects. Muscle acetylcholine receptors are ion channel receptors—receptors that form pores in response to extracellular signals, allowing ions to flow in or out of cells.[6] In the absence of acetylcholine, five subunits of the acetylcholine receptor, each with four transmembrane barrels, cluster together (Figure 7.3). When acetylcholine binds to these receptors, the subunits contort, moving apart to form an ion channel. Positively charged ions flow into muscle cells through these channels, increasing membrane voltage.[7]

Acetylcholine, in summary, generates a change in membrane voltage by opening ion channels in a receiving cell. Notice that

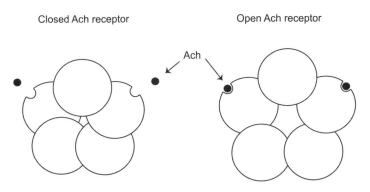

Closed Ach receptor   Open Ach receptor

Ach

7.3. Acetylcholine (Ach) receptor ion channel structure. Cross-sections through closed (left) and open (right) acetylcholine receptor ion channels are depicted. Small dots represent acetylcholine molecules.

these ion channels differ from the voltage-sensitive and stretch-sensitive channels introduced in Chapter 6, responding to a neurotransmitter rather than membrane voltage or stretch. Acetylcholine receptors are members of a large class of ion channels that respond to neurotransmitters, hormones, and other chemicals.

Let's face it: some kisses stoke your fire; some don't. The same is true of synapses: some excite; some inhibit. Action potentials fire when membrane voltage increases above a threshold, opening voltage-sensitive sodium channels. Excitatory synapses promote action potentials in receiving cells by increasing membrane voltage toward the threshold. Inhibitory synapses suppress action potentials by decreasing voltage away from the threshold.

The neuromuscular junction excites muscle cells. Approximately 150,000 acetylcholine receptors open in response to each motor neuron action potential, generating large influxes of sodium.[8] This excitation reliably brings voltage above the action potential threshold in patches of membrane near the neuromuscular junction. The result is one-to-one transmission of action potentials between motor neurons and muscle cells. Neuromuscular disorders can disrupt the fidelity of this communication. For example, patients with myasthenia gravis have fewer acetylcholine receptors, resulting in blunted responses to acetylcholine. Thus stimulation by motor neurons sometimes does not activate action potentials. As a result, muscles fail to contract in response to neural signals, leading to weakness.

Why bother with all this complex machinery simply to pass an action potential in one cell along to another cell? Why don't muscle and nerve cell membranes fuse, creating direct electrical connections? One explanation is that information flows one way at the neuromuscular junction—from nerve to muscle. Chemical synapses prevent the chaos that might result if action potentials flowed backward from muscles to nerves. But we need to look beyond the atypically simple one-to-one connection at the neuromuscular junction to appreciate the most crucial attributes of chemical synapses: their variability, combinability, and flexibility. Synapses have diverse effects, they collaborate with each other, and they are malleable, subject to both short- and long-term modulation.

Suppose it were your job to adjust the water temperature of a busy swimming pool. Wishing to be responsive to the swimmers, you occasionally get everyone's attention and ask whether to turn the heat up or down. Responses, of course, will vary. Some folks will want to be warmer, others cooler. Passionate folks will scream, indifferent folks will whisper. How will you sort the responses? Will everyone's opinion be taken equally? Will regulars be favored over first-timers? Will nearby folks influence you more than those across the pool? Most neurons face similar choices as they weigh input from various sources and make decisions.

Synapses from sensory neurons, interneurons, and the brain—some excitatory, some inhibitory—converge on the cell body and dendrites of motor neurons.[9] Motor neurons synthesize this cacophony of inputs into an output: the frequency of action potentials firing down their axons. The mechanism is simple. Changes in membrane voltage flow from synapses to the action-potential-initiating zone near the junction of axon and cell body, a place where voltage-sensitive sodium channels reside at high density. At rest, the initiating zone membrane voltage is below threshold for action potentials, and the motor neuron is silent. When voltage climbs above threshold, voltage-sensitive channels open, action potentials fire, and muscles contract. As membrane voltage rises farther above threshold, firing frequency increases, producing progressively stronger muscle contractions.

At most excitatory synapses onto motor neurons, sending cells release the neurotransmitter glutamate, which binds to glutamate receptors on receiving cells. Much like the acetylcholine receptors on muscle cells, these glutamate receptors are ion channel receptors assembled from multiple subunits. When they bind glutamate, these receptors open and positive charge flows into the cell, producing excitation. At inhibitory synapses, neurotransmitters bind to receptors that allow negatively charged chloride ions to flow into neurons, causing membrane voltage to become more negative.[10] Inhibitory synapses thus oppose the actions of excitatory ones.

The response of motor neurons to synaptic input ultimately depends on the integration of excitatory and inhibitory inputs. Each synapse's influence depends on its strength and position. Synaptic strength—the amount of voltage change a synapse produces in receiving cells—depends on the amount of neurotransmitter released by sending cells, the properties of receptors on receiving cells, and other factors.[11] Synapses that are distant from the initiating zone exert small influences because synaptic potentials decay as they propagate. So just as we are more likely to be influenced by a nearby strident voice than a distant weak one, strong synapses close to the axon's initiating zone have the greatest influence on motor neuron output.

The language of logic—such as IF-THEN statements and AND, OR, and NOT operators—can help us envision how neurons make decisions. The following statement accurately describes the simple communication at the neuromuscular junction:

> IF the motor neuron fires action potentials, THEN the muscle fires an action potential.

More complex decision making becomes possible with multiple inputs. For example:

> IF a neuron receives excitatory input 1 AND excitatory input 2, THEN it fires an action potential.

In this case, the neuron reaches threshold only if two excitatory inputs summate. Alternatively:

> IF a neuron receives excitatory input 1 OR excitatory input 2, THEN it fires an action potential.

In this instance, excitation from either synapse suffices to bring the neuron above threshold. Finally, inhibitory inputs make NOT operations possible:

> IF a neuron receives excitatory input 1 but does NOT receive inhibitory input 1, THEN it fires an action potential.

In this situation, excitatory input 1 brings the neuron above threshold unless inhibitory input 1 is also active. With many thousand inputs into each motor neuron and the possibility of combining the AND, OR, and NOT operations, the opportunities for making complex decisions exist.

These theoretical scenarios have real-world parallels in motor neurons, whose cell bodies reside in the spinal cord. Axons that descend from the brain to the spinal cord excite motor neurons through direct connections and by activating spinal interneurons. Sensory neurons that ascend from tissues to the spinal cord, such as those connected to muscle stretch receptors, also excite motor neurons. We can envision AND, OR, and NOT operations at work in these inputs to motor neurons.

Each motor neuron receives multiple inputs from the brain. In many cases, activating a motor neuron requires input from several of these pathways—an example of the AND operation. Action potential thresholds differ across motor neurons: those that innervate slow-twitch fibers have lower thresholds than those that innervate fast-twitch fibers. So slow-twitch motor units activate in response to modest excitatory input, whereas fast-twitch motor units require further excitatory input to fire. In other words, fast-twitch muscle fibers need the input required to activate slow-twitch fibers AND additional inputs. When our muscles contract, slow-twitch fibers always recruit first because the slow-twitch motor neurons reach threshold before those of fast-twitch fibers. Low-intensity contractions thus use only slow-twitch fibers, allowing for sustained energy-efficient efforts. Fast-twitch fibers join only during powerful contractions, saving these fatigable fibers for the times when we need their contributions.

The interaction of ascending and descending inputs illustrates an OR operation. During a voluntary contraction, commands descending from the brain can activate the quadriceps without assistance of ascending input. During the knee-jerk reflex, sensory signals ascending from the muscle spindle can activate the quadriceps—in this case, no descending signals from the brain are necessary. In other words, strong descending OR powerful ascending input can trigger quadriceps contractions.

Try this experiment: while seated, place both feet flat on the ground and contract your right calf maximally without moving your leg or foot. It might take a few tries to learn to make these isometric contractions. Repeat on the left side. Finally, make a maximal bilateral contraction—simultaneously contract both calves. Do you notice a difference between unilateral and bilateral contractions? In most people, contraction strength declines when both legs contract simultaneously. This bilateral deficit may be the result of inhibitory interneurons that project between the left and right sides of the spinal cord: exciting the right calf activates an inhibitory interneuron that opposes excitation of the left calf. This circuitry probably helps produce the alternating contractions required for bipedal locomotion—the left calf must relax while the right calf contracts. However, the circuitry somewhat weakens bilateral efforts because contractions on each side of the body inhibit those on the other.[12]

Bilateral contractions exemplify integration of excitatory and inhibitory inputs. Inhibitory signals from the opposite limb reduce, but don't prevent, the increase in membrane voltage caused by excitatory input. For some motor units, the inhibitory input may act like a NOT command—for example, fast-twitch motor units may not reach threshold and therefore not activate at all. At most motor units, the inhibition is probably more like twisting the volume knob, decreasing action potential frequency. So although muscles still activate during bilateral contractions, force production declines. Training can apparently override the bilateral deficit. Weight lifters, rowers, and other athletes who regularly make bilateral contractions may instead show bilateral facilitation—stronger contractions when making simultaneous efforts with both limbs. These athletes probably produce enough excitatory stimulation to supersede the inhibitory signals.[13]

"I went out too fast" is the most common lament at postmarathon parties. Despite countless miles at race pace during training, despite obsessive prerace pace planning, despite the exhortations of every running guru, runners still go foolishly fast at the beginning of marathons. Unfortunately, dire consequences await those who start too fast: the marathon offers an extended opportunity to suffer. How could so many runners—including experienced ones—make the same dumb mistake? One problem is that races mess with our perception of pace. When the gun goes off, we are running faster than we think: at the front of the pack five-minute miles feel like sixes, toward the back tens feel like twelves.

Modulatory synaptic input to motor neurons may partly explain the recalibration of our internal pace meters at the start of races. Like all membrane proteins, receptors at modulatory synapses have extracellular, transmembrane, and intracellular domains. But modulatory receptors differ from ion channel receptors. Rather than simply providing a path for ions to flow, they morph when neurotransmitters bind to their extracellular domains, altering the shape of their intracellular domains. The morphed intracellular domains interact differently with cellular proteins, triggering signaling pathways that ultimately regulate kinases and phosphatases (Chapter 5). Recall that kinases and phosphatases add and remove phosphates from target proteins, resulting in sustained changes in function. These effects of modulatory synapses take longer to develop but last longer than those mediated by ion-channel receptors. Modulatory synapses typically alter how neurons respond to other synaptic input, often by modifying ion-channel properties.

Fear, anxiety, and excitement activate modulatory synapses on motor neurons that release noradrenalin, a neurotransmitter related to the hormone adrenalin.[14] At these synapses, noradrenalin morphs ion channels. But rather than influencing ion channels directly associated with excitatory synapses, it alters ancillary channels, making them more sensitive to changes in membrane voltage. When membrane voltage increases due to ion-channel activation at excitatory synapses, these sensitized channels open too. Thus noradrenalin amplifies each excitatory input because it causes additional channels to open, resulting in bigger synaptic potentials. So noradrenalin doesn't itself excite motor neurons, but instead increases their *excitability,* making them more responsive to excitatory input.[15]

The possible link to our perception of race pace may be that noradrenalin input to motor neurons comes from brainstem regions that respond to our level of arousal. The systems that release noradrenaline are quiet during sleep, which may explain why it's hard to exercise right after awakening—there is little amplification of excitatory signaling. In contrast, fear and anxiety dramatically stimulate the noradrenalin system—and everyone is afraid and anxious at the start of a marathon. Because noradrenalin amplifies excitation, the same stimulus may produce stronger muscle contractions during a marathon compared to a training run. Thus racers may need to deliberately pull back the reins or risk an agonizing shuffle to the finish.[16]

Just before starting this book, I had ankle surgery that prevented me from running for four months. During this layoff, my friends were remarkably kind, regularly checking in on my recovery. Their concerns, I began to realize, were not only about my physical health, but also about my psychological well-being. "What's it like being an addict denied your fix?," I heard them asking. Of course, no one asked it quite that bluntly, and maybe my own anxieties caused me to hear that question even when it wasn't intended. In any case, the worries were well founded. Exercise triggers the same neural pathways as gambling, fatty foods, and certain illegal drugs, stimulating pleasure circuits that encourage addiction.[17] Exercise, some might argue, is a cheap, safe alternative to self-destructive behaviors. Or perhaps put more accurately, many of our vices are dangerous shortcuts to the natural high of exercise.

Like the motor coordination circuitry, the pathways that produce pleasurable sensations span multiple brain regions. At the core of these paths are "pleasure neurons" with cell bodies in a part of the brain called the ventral tegmental area. Pleasure neurons send axons to another region, the nucleus accumbens, where they release the neurotransmitter dopamine.[18] And, to make a long story short, elevated dopamine in the nucleus accumbens feels good. Many drugs of abuse—as well as other addictive behaviors—act upon this dopamine system.

Although distance runners typically go easy on recreational drugs, they may nonetheless seek the same euphoria as pot smokers and dope addicts. The chemicals that produce runner's high— cannabinoids and opioids—resemble the active ingredients of marijuana and heroin. Our bodies synthesize cannabinoids and opioids that stimulate dopamine release by pleasure neurons. Rather than acting directly on pleasure neurons, these molecules bind to receptors on sending cells that make inhibitory synapses onto pleasure neurons. When activated, opioid and cannabinoid receptors block release of inhibitory neurotransmitters from these sending cells, reducing inhibitory input to pleasure neurons. So, the opioid and cannabinoid receptors are modulatory—they hamper inhibitory synapses, thereby exciting pleasure neurons to release dopamine.

For decades, scientists attributed runner's high to opioids, because blood levels of the opioid beta-endorphin rise during exercise. But this hypothesis had a major flaw: beta-endorphins don't cross the blood-brain barrier, so blood beta-endorphins don't affect our pleasure circuitry. A recent study, however, has revived the opioid hypothesis. Using brain scans, scientists measured increased opioids in the brains of runners after a two-hour workout. Thus exercise apparently stimulates opioid synthesis inside the brain. Postworkout euphoria correlated highly with brain opioid level, bolstering the argument that these chemicals contribute to runner's high.[19]

Tantalizing evidence also implicates cannabinoids in runner's high. Cannabinoids definitely rise during exercise, exerting wide influence. During exercise, blood cannabinoids cause increased blood flow in some tissues. Moreover, portions of the brain that control movement release cannabinoids. These naturally elevated cannabinoids are plausible causes of runner's high because the characteristics of runner's high—elation, pain-relief, distorted perception, feelings of harmony—resemble the effects of recreational drugs that contain cannabinoids.

Disruptions to cannabinoid signaling apparently reduce the desire to run. In one study, normal rats voluntarily ran about twenty-five miles per week on an exercise wheel—about the same weekly mileage as a typical human recreational runner!

Rats with cannabinoid receptor knockouts, in contrast, ran only about fifteen miles per week, suggesting that cannabinoid signaling motivates running behavior. However, the jury is still out on cannabinoids' role in runner's high. Mice without cannabinoid receptors may have a reduced capacity to run, as well as a reduced desire, because of the receptors' role in motor coordination. And studies using inhibitors to block cannabinoid signaling have produced mixed results.[20]

On the day I wrote this section, I also ran a speed workout. (My ankle is healing!) To be honest, I felt a great deal more distress than ecstasy. Toward the end of each 400-meter repeat, slow waves of nausea rode atop acute breathlessness. But perhaps this explains why exercise stimulates our pleasure pathways. Exercise hurts, especially when we push hard. Without some mechanism to counter this pain, we might not make the effort. For me and most other modern humans, the effects of inactivity would be negative, but not immediately life threatening. For our ancestors, however, staying alive probably required strenuous physical activity. Hunting prey, gathering food, avoiding predators, and battling other humans all demand vigorous exercise. Runner's high, in other words, might be a basic survival mechanism.

Although the warm afterglow of a workout fades after a few hours, other effects of exercise last much longer. When we practice shooting free throws, kicking penalty shots, or swimming efficiently, the improvements persist. And we never forget some motor skills—such as riding a bicycle. Practice improves performance by molding neural circuitry through sustained modifications to synapses. A simple principle guides this process: use it or lose it. When we practice, we reinforce the synapses along the activated circuits, fortifying the circuitry. Synapses that originally whispered begin screaming, whereas unused synapses wither and perish. Practiced movements, activated by busy neural highways, are precise, efficient, and automatic. Unpracticed movements, activated by obscure neural footpaths, are clumsy and uncoordinated.

How does practice modify neural pathways? One mechanism is long-term potentiation (LTP), the sustained strengthening of excitatory synapses. To envision how LTP works, imagine yourself on a crowded train. Suppose a fellow passenger elbows you in the back. How do you respond? Unless you are in a particularly foul mood, you will probably show some restraint. Maybe the push was an accident. Now, suppose that same passenger elbows you every day for a week. How do you respond to this repeated abuse? One possibility: you accept the occasional bump as part of the train experience and respond even less forcefully. Synapses sometimes demonstrate such tolerant behavior—they desensitize, thereby requiring bigger stimuli before activating. Another possibility is that your patience wears thin and you respond more forcefully to the constant harassment, perhaps throwing an elbow of your own. This response resembles potentiation of synapses— the strengthening of synaptic strength in response to repeated stimuli. Now suppose a passenger bumps you every day for a week, leaves you alone for a month, and then bumps you again. If you remember the original affronts and therefore respond vigorously, your response resembles LTP.

Scientists demonstrate LTP by putting excitatory synapses through an ordeal similar to the train scenario just described: first, they activate the sending cell of a synapse and measure synaptic strength by quantifying the resultant change in voltage of the receiving cell's membrane. This test stimulus resembles the first bump. Then, after stimulating the synapse at high frequency (analogous to the daily bumps), scientists again measure synaptic strength in response to a single test stimulus. By varying the time between the high-frequency stimulation and the second measurement of synaptic strength, scientists can learn how long synapses "remember" the vigorous stimulation. In some synapses, forceful stimulation provokes persistent increases in synaptic strength. In other words, these pugnacious synapses seem to remember powerful activation, responding more forcefully for hours afterwards. Because LTP produces long-lasting changes, it can power both learning and memory.

One type of glutamate receptor—the N-methyl-D-aspartate (NMDA) receptor—initiates LTP, translating strong activation into sustained synaptic remodeling. These receptors play hard-to-get. Although most channels respond to either chemicals or voltage, NMDA-receptor ion channels need both the chemical kiss of glutamate *and* the electrical charge of increased membrane voltage. So before these coy receptors open, other glutamate receptors must first produce excitatory potentials. NMDA receptors thus track synaptic traffic, because they activate only when synapses are already stimulated. Open NMDA-receptor channels allow sodium and calcium to flow into cells. This influx of positive charge augments the excitatory stimulus. Also, calcium triggers lasting effects in receiving cells, modifying synapse architecture. Many synapses reside on small spines projecting from dendrites. These spines widen in response to repeated stimulation, resulting in a larger surface area that fills with newly synthesized receptors. Physical connections also form when proteins anchored in membranes of both sending and receiving neurons zip together to link the neurons on opposite sides of a synapse.[21]

LTP has a central role in sculpting our brain-body connections. When we learn to juggle, throw strikes, or nail three-point baskets, the synapses we activate respond like muscles—they bulk up. In other words, practice not only strengthens our muscles but also the neural circuits that stimulate them.

The long-term effects of exercise on our movement circuitry underpin the incomparable thrill associated with learning a new sport and the deep satisfaction of improving through practice. But they are not that surprising. "Use it or lose it" makes sense; we expect to get better when we practice something, whether it's chess, badminton, or the breaststroke. On the other hand, the broader effects of exercise on the brain function are astonishing. Physical activity forestalls depression, improves memory and learning, and offsets declines in brainpower that accompany disease and aging. So forget about the "dumb-jock" stereotype—physical and mental fitness go hand in hand.

When we do biceps curls, we don't expect to strengthen our calves. Yet exercise provokes widespread brain benefits, extending far beyond the movement circuitry. The hippocampus—a part of the cerebral cortex with leading roles in learning and memory—responds vigorously to exercise. Recent work shows that a year of aerobic exercise expands the hippocampus of older adults by 2 percent, offsetting age-related declines.[22] Exercise renovates the brain by stimulating growth factors, proteins that foster cell proliferation and remodeling. Cells outside the brain release insulin-like growth factor (IGF) and vascular endothelial growth factor (VEGF), which both cross the blood-brain barrier and stimulate brain cells. These factors also contribute to changes in muscles during exercise training (see Chapter 11). Cells inside the brain, including those in the hippocampus, produce brain-derived neurotrophic factor (BDNF).[23]

Neurons blossom when sprinkled with BDNF. BDNF triggers neurogenesis—the sprouting of new neurons from stem cells. Moreover, BDNF offsets neurodegeneration—the wilting of neurons resulting from aging, anxiety, and ailments. And BDNF promotes LTP—the key to memory and learning. These effects have led psychiatrist Dr. John J. Ratey to dub BDNF "Miracle Gro for the brain."[24] Exercise fertilizes neurons, not only fostering their growth but also making them more resilient, that is, better able to cope with distress and disease.

BDNF binds to tropomyosin receptor kinase B (TrkB), a modulatory receptor that is also a kinase. Upon BDNF binding, lone TrkB receptors pair up and add phosphate groups to their partners, morphing them into active form. These activated receptors then trigger a host of signaling pathways, ultimately leading to synthesis of new proteins and long-lasting effects. Both sending and receiving cells express TrkB receptors, so BDNF signaling lacks the restrictive one-way pattern of most synaptic communication. BDNF released from receiving cells, for example, can flow backward and bind to receptors on sending cells. Consequently, BDNF signaling influences both sides of the synapse, accounting for its wide-ranging effects.[25]

At the start of this chapter, I described a simple linear model for controlling movement. Our initial picture featured reliable, fixed, one-way, one-to-one connections. The rest of the chapter has systematically complicated the picture. To understand motor coordination, we need to contemplate the convergence and divergence of hundreds, even thousands, of connections. We must accept conditional, qualified signaling and view synapses as plastic, flexible, and malleable. And we cannot forget the feedback between brain and body. So rather than envisioning our bossy brains dictating instructions to our servant bodies, we need to appreciate the dynamic interplay between brain and body, to welcome the body's role in modulating the brain's commands, and to embrace the power of exercise to mold our brains.

A bottom-line message, one that applies to every part of exercise physiology, is to beware the overly simplistic explanation. Or, to put it in positive terms, savor the complexity beneath the elegance of humans in motion.

Chapter 8

Live High

I do LSD most Sunday mornings. For me and other runners, LSD stands for "long slow distance"—the keystone of endurance training. A typical LSD run covers sixteen to twenty miles and takes about three hours. When I'm running alone, LSD often mesmerizes me. The calming cadence of my legs, lungs, and heart overtakes my awareness. In my semiconscious state, I sometimes visualize the molecular tempo too: the waves of calcium that trigger heart contractions, the pivoting of myosin that produces muscular force, and the pulsations of hemoglobin that deliver oxygen to hungry muscles.

Max Perutz recognized the similarities between whole-body and molecular rhythms in the 1950s. Perutz, who determined hemoglobin's structure, dubbed it a "molecular lung."[1] His point was not only that hemoglobin, like lungs, helps carry oxygen from one place to another. He also meant to highlight the similarity of the cyclical thumping of our hearts and lungs to the throbbing of hemoglobin as it binds and releases oxygen. Hemoglobin is not a passive oxygen-binding protein, but rather a dynamic shape-shifter, expanding and contracting during its journey through the body. Hemoglobin plays a leading role in aerobic metabolism, the topic of this chapter.

Compare aerobic metabolism to a fire in your fireplace. Open the flue, crumple some newspapers, and light them. Energy stored in the paper's chemical bonds quickly converts to heat and light, while oxygen rushes into the fireplace to fuel the fire. After the blaze, a small pile of ashes remains, weighing much less than the starting material. The fire has transformed organic material into carbon dioxide and water vapor, which have escaped up the chimney. Likewise, aerobic metabolism burns sugar (or fat), consuming oxygen and releasing carbon dioxide, water, and energy.

Though aerobic metabolism resembles fire, key differences exist. First, neither sugar nor fat burn as readily as newspaper or wood—try lighting a pile of table sugar or a stick of butter with a match. Also, fires produce light and heat, which are cozy but can't be put to further use. Aerobic metabolism, in contrast, transfers the chemical energy of foods into adenosine triphosphate (ATP), which can power cellular processes. Finally, rather than happening all in one place, aerobic metabolism occurs throughout the body. Each cell tends its own tiny fire, extracting only the energy it immediately needs. So rather than a single, all-out blaze of highly flammable substances, aerobic metabolism is a dispersed, stepwise, controlled burning of modestly combustible materials.

Aerobic metabolism obtains energy from nutrients by stripping away their electrons. Negatively charged electrons occupy the periphery of atoms, swirling around the uncharged neutrons and

positively charged protons that form the nucleus. The most energized electrons circle farthest from the nucleus, poised to release
energy if they succumb to the nucleus's pull and fall toward it.
Energy-rich nutrients such as glucose and fat contain many high-
energy electrons. Metabolic pathways transfer these electrons to
other molecules, extracting some of the electrons' energy. High-
energy electrons are like full squeeze bottles, the kind you see on
the sidelines of a football game. In this analogy, the energy drink
inside a squeeze bottle corresponds to an electron's energy content.
Players drink from bottles and then pass them along to teammates.
The bottles progressively empty and the players use the ingested
energy to power their muscle contractions. In the same way, some
metabolic reactions transfer electrons from molecule to molecule.
With each move, the electrons lose some energy. Aerobic metabolism captures some of this energy and spends it to make ATP.

Metabolic pathways can convert both sugar and fat to acetate, a
molecule that contains many high-energy electrons. Acetate enters the citric acid cycle, where a series of reactions progressively
picks it apart, releasing carbon dioxide and high-energy electrons
(Chapter 5).[2] During the stepwise breakdown of acetate, electron
carriers harvest the released electrons. Electron carriers—molecules
that readily bind and release electrons—resemble the trainers
who carry bottles out to the huddle where players consume their
energy-rich contents. During metabolism, electron carriers trap
and transport electrons.

The citric acid cycle occurs in mitochondria, intracellular structures with a double membrane (Figure 8.1). Electron carriers
convey energized electrons to proteins and other molecules on
the inner mitochondrial membrane. Energized electrons jump
from one mitochondrial membrane molecule to the next, pro-

gressively giving up their energy, in the same way that a squeeze bottle empties as players hand it around the huddle. The last molecule in the sequence expels energy-depleted electrons. Oxygen mops up this mess by reacting with used-up electrons, combining with them and protons (H^+ ions) to form water.[3] In the absence of oxygen, these electrons lack a final resting place; so they build up, causing aerobic metabolism to grind to a halt.

At the end of this chapter, we will explore how mitochondria use the energy stripped from electrons to synthesize ATP. But first, let's look closely at oxygen transport. Tissues need to receive exactly enough oxygen to meet their metabolic demands. Too much oxygen damages deoxyribonucleic acid (DNA) and proteins, whereas too little oxygen impairs aerobic metabolism. Because our muscles consume up to seventy-fold more oxygen during exercise than rest, we must carefully control the rate of oxygen delivery.[4] Increased blood flow accounts for much of the enhanced oxygen delivery to muscles during exercise; the remarkable

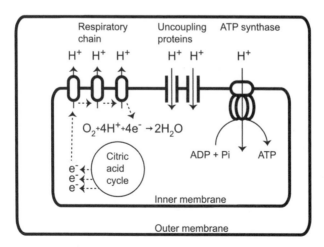

8.1. Structure of mitochondria. Dotted lines indicate the path of electrons.

properties of the oxygen-binding protein hemoglobin account for most of the rest.[5]

Oxygen abounds in the air around us, but delivering it to exercising muscles poses problems. To understand these challenges, let's consider a much simpler situation—an open beaker of water. Scientists describe the amount of oxygen in the air above the beaker as its partial pressure, the contribution of oxygen to the total air pressure. At sea level, air pressure is 760 torr, and oxygen is about 21 percent of the total, so the oxygen partial pressure (P_{oxygen}) is about 160 torr.[6] If the beaker's water initially contains no oxygen, a large gradient exists for diffusion of oxygen into the water. Thus oxygen diffuses into the beaker and the beaker's P_{oxygen} eventually rises to match the air's P_{oxygen}.

During exercise, oxygen travels from the air to our muscle cells, where P_{oxygen} ranges from 10–40 torr. Similar to the beaker example, a large gradient exists for oxygen diffusion from atmosphere to muscles. But there's a problem: our bodies are watery, and oxygen diffuses very slowly through water, about 10,000 times more slowly than through air. For this reason, only the very smallest organisms—those less than a millimeter in diameter—rely solely on diffusion for oxygen transport. Humans and other large creatures must use bulk flow, the movement of substances along with currents of gas or liquid, to move oxygen across the large distances between the atmosphere and our muscles. Bulk flow carries everything along with the flow of air or water, whereas diffusion moves specific molecules according to their own concentration gradients.[7]

Recall from Chapter 1 that blood pressure gradients drive flow through the circulatory system. This bulk flow of blood carries oxygen molecules through blood vessels like corks swept down a river. Similarly, differences in total air pressure move air into and out of our lungs by bulk flow, moving oxygen into our lungs like leaves blown by wind. In both air and blood, oxygen gets to its destination much faster by bulk flow than it would by diffusion alone.

Oxygen's voyage—its trip from the atmosphere to our cells—follows a complex itinerary, alternating between bulk flow and diffusion. When our chest cavity expands, total air pressure inside the lungs falls, and air rushes in by bulk flow. Oxygen then diffuses from the lung into the blood, which carries it by bulk flow to the muscles. Finally, oxygen diffuses from the blood to muscles. The pattern resembles a typical plane trip, where the flights are analogous to bulk flow, quickly carrying people en masse over long distances. Everybody on a plane goes to the same destination, just as blood flows around our circulatory system carrying all its components—including oxygen and carbon dioxide—in the same direction. After flights, some people saunter to the next gate, whereas others go to the taxi stand, but in both cases they are moving much more slowly than a plane. These walks resemble diffusion because they are relatively slow and cover small distances, and because individuals go their own ways. Similarly, oxygen and carbon dioxide diffuse in opposite directions along their own concentration gradients. Oxygen flows from lung to blood to muscle, and carbon dioxide flows in the opposite direction because of these differences in the direction of their diffusion, just as the different places where people walk after a flight determines their final destinations.

Bulk flow speeds oxygen transport by whisking it quickly across long stretches of its journey, greatly shortening the distance it needs to diffuse. But there's another problem: oxygen doesn't dissolve well in water. When water comes to the same P_{oxygen} as air, it contains twenty to forty times less oxygen. For this reason, if water flowed through our arteries, not nearly enough oxygen would be carried to the tissues. Fortunately blood, as we all know, is thicker than water. A cup of blood contains about a trillion red cells, each packed with about 250 million hemoglobin molecules.[8] At 100 torr, the lung's oxygen level, blood holds sixty-fold more oxygen than water, 98 percent of it bound to hemoglobin.

Hemoglobin delivers just the right amount of oxygen to muscles because it responds to the surrounding conditions. Human hemoglobin contains four globular protein subunits, each cradling a non-protein heme group, a disclike molecule with an iron atom at its center.[9] When oxygen binds to these iron atoms, the heme groups morph from a slightly buckled disc to a flat plane. Though subtle, these heme group contortions transmit to the protein chains and affect interactions between subunits. In the oxygen-depleted state, hemoglobin tenses—the subunits bind tightly to each other, restricting oxygen binding to heme. But oxygen binding to one or more subunits relaxes the entire hemoglobin, loosening connections between subunits and granting oxygen better access to its binding sites. As a result, oxygen binds hemoglobin cooperatively: the entire hemoglobin molecule morphs when oxygen attaches to one subunit, enhancing oxygen binding to other subunits. So hemoglobin subunits are the ultimate team players. They contribute not only by doing their own oxygen-binding work, but also by enabling their teammate subunits to better bind oxygen.[10]

Scientists measure hemoglobin's teamwork by exposing hemoglobin-rich red blood cells to a range of oxygen partial pressures. Hemoglobin's heme groups glow bright red when oxygen binds but dim to a more subtle purplish-red in the absence of oxygen, so scientists can evaluate oxygen binding by monitoring hemoglobin's color. Graphing the amount of oxygen bound to hemoglobin at different environmental oxygen levels reveals hemoglobin's remarkable features (Figure 8.2). In the absence of oxygen, no oxygen binds to the tense hemoglobin molecules. As P_{oxygen} increases, oxygen weakly binds the tense hemoglobin, and the binding curve slopes upward gently. However, as more oxygen binds, hemoglobin transitions to the relaxed state, making further oxygen binding easier. As a result, hemoglobin binding increases steeply over a small range of P_{oxygen}. At higher P_{oxygen}, hemoglobin saturates with oxygen, leaving no binding sites for additional oxygen, so the binding curve flattens.

To oxygenate muscles, hemoglobin must bind oxygen as blood flows through the lungs and release it as blood flows past muscles.

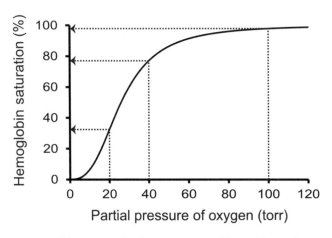

8.2. Hemoglobin oxygen-binding curve. Dotted lines indicate the partial pressure of oxygen in lung air (100 torr), resting muscles (40 torr), and exercising muscles (20 torr).

Binding at the lungs—the loading zone—depends upon lung oxygen levels, which can fall when we go to altitude, hold our breath, or start using more oxygen during exercise. But notice that the hemoglobin oxygen-binding curve flattens between 80 and 120 torr P_{oxygen}. Hemoglobin, for example, remains 95 percent saturated when P_{oxygen} falls to 80 torr. Consequently, small variations from the normal lung P_{oxygen} of 100 torr have little effect on the amount of oxygen carried away from the lungs by hemoglobin. For this reason, holding your breath for a moment does not substantially impair oxygen delivery to tissues because even as lung P_{oxygen} drops a bit, hemoglobin remains almost fully saturated with oxygen. Hemoglobin promotes homeostasis by maintaining rather constant blood oxygenation even when lung oxygen levels fluctuate.

In contrast, the hemoglobin oxygen-binding curve steepens at the unloading zone. In resting muscles, P_{oxygen} is 40 torr, and hemoglobin remains almost 80 percent saturated with oxygen. So hemoglobin unloads only about 20 percent of its oxygen when it flows past a resting muscle. This seems like an inefficient system—why lug all that oxygen down to the muscles then back to the lungs? But the benefits become clear during exercise. When exercising muscles rapidly use oxygen, their P_{oxygen} drops to about 20 torr, causing the oxygen saturation of hemoglobin to fall to about 30 percent. Consequently, hemoglobin dumps about 70 percent of its oxygen when it passes an exercising muscle. Hemoglobin's cooperativity—the teamwork between hemoglobin subunits—sensitizes it to changes in muscle P_{oxygen}, so hemoglobin delivers three-fold more oxygen to needy exercising muscles compared to resting ones.

Hemoglobin not only hauls oxygen from lungs to muscles, but also carts carbon dioxide the opposite direction, capturing it at the muscles and releasing it at the lungs. We might envision oxygen and carbon dioxide as two suitors for hemoglobin's attention. Carbon dioxide and oxygen both tend to kick each other off hemoglobin, almost as if they want hemoglobin all to themselves. At the tissues, where carbon dioxide levels are high, binding of carbon dioxide to hemoglobin morphs it toward the tense form, increasing the release of oxygen. This effect is particularly pronounced in exercising muscles, where carbon dioxide levels rise as a result of active metabolism. H^+ ions also rise during exercise, and they too morph hemoglobin toward the tense state, further increasing oxygen release. In other words, local accumulations of carbon dioxide and H^+ ions prod hemoglobin to dump oxygen, enhancing delivery of needed oxygen to exercising muscles.

The bottom line is that the hemoglobin molecule itself, without any intervention by hormones or nerves, regulates oxygen delivery to tissues. Hemoglobin contorts in response to oxygen, carbon dioxide, and H^+ levels, automatically dumping more oxygen at exercising muscles, thereby helping to meet their increased demand.

Red blood cells live a short, tough life. With each circuit around the body, they squeeze through the miniscule blood vessels of a capillary bed. Red cells deform as they pass through these tiny conduits. The tight fit maximizes gas and nutrient exchange between red cells and tissues, but it also batters red cells, shortening their lifespan to about 120 days. Because the lifespan of red cells is brief, we must constantly synthesize them—our bone marrow makes about a million red cells every second.[11]

Unsurprisingly, our bodies tightly regulate red cell synthesis. Red cells normally are about 40–45 percent of blood volume. Men tend to have somewhat higher red-cell levels than women, in part because testosterone stimulates red-cell synthesis. Anemia—when red-cell levels fall below 35 percent—diminishes blood's oxygen-carrying capacity. Mild anemia harms athletic performance; severe anemia is life threatening. On the other hand, blood thickens and becomes viscous when red-cell levels rise above about 55 percent. As a result, blood pressure surges, increasing the risk of heart failure and stroke.

Athletes' aerobic capacity, their maximal rate of oxygen uptake (VO_{2max}), depends heavily on their blood's red-cell concentration. Any step along oxygen's odyssey—its trip through the lungs, its journey through the blood, or its consumption by cells—could theoretically limit VO_{2max}. But under most conditions, blood oxygen transport probably limits oxygen uptake.[12] Boosting red cell levels elevates VO_{2max} by increasing the amount of oxygen in each drop of blood. Because VO_{2max} correlates highly with endurance, distance athletes often seek to increase their red-cell levels.

Athletes have long been tempted to take shortcuts. Blood doping—the illicit, artificial expansion of red-cell levels—comes in two flavors. Cheaters can transfuse blood: they harvest their own red cells many weeks before a competition, store the cells, and then reinfuse them just before the event. During the interim between withdrawal and infusion, red-cell levels in the blood rebound, so the infusion expands their levels above normal. Dishonest athletes can also inject erythropoietin (EPO). This protein hormone, which is released by kidney cells in response to low oxygen levels, binds to receptors on red-cell precursors in the bone marrow. Without EPO, these precursors die. EPO stimulates the precursor cells to thrive and divide, increasing red-cell production. Doping carries risks, including heart failure due to dangerously high red-cell concentrations. Though both methods of doping increase an athlete's own red cells, tests can detect subtle differences between doping and natural increases, thereby catching cheaters.[13]

Some athletes don't need doping to attain high red-cell counts. EPO works by binding to receptors on membranes of red-cell precursors. Members of a Finnish family carry one EPO receptor allele with a rare nonsense mutation (see Chapter 3) that truncates their EPO receptors. The missing portion of the receptor normally serves as a brake, slowing the receptor's activity. Lacking this inhibition, the truncated receptor responds especially aggressively to EPO, leading to increases in red-cell and hemoglobin levels. One member of the Finnish family won three gold medals in cross-country skiing at the 1964 Winter Olympics. Although the EPO receptor mutation appeared to enhance endurance in this individual, others with truncated EPO receptors experience negative consequences, including cardiovascular disorders.[14]

Honest athletes who lack mutations in their EPO receptors can try to expand their red-cell count by going to high altitude. Altitude stimulates red-cell synthesis because hypoxia—low oxygen levels—naturally increases blood EPO concentration. How does an environmental factor such as hypoxia affect the production of EPO? To address this question, we need to consider the regulation of genes. Each cell has a complete set of DNA, containing every gene in the genome. Yet each cell activates only a fraction of these genes, so at any given time and location, many genes remain quiescent and do not produce proteins. And cells tightly regulate the activity of some genes—such as the one coding for EPO—rapidly turning them on or off.

The cell's nucleus sequesters DNA, which contains the instructions for making proteins, much as a library might store blueprints for building houses (Figure 8.3). Outside the nucleus, factories called ribosomes synthesize proteins. Ribonucleic acid (RNA) ferries instructions for protein synthesis from the DNA to the ribosome. Inside the nucleus, enzymes transcribe DNA into RNA. Outside the nucleus, ribosomes translate RNA into protein. The rates of RNA and protein synthesis, as well as their degradation rates, determine the abundance of each protein.

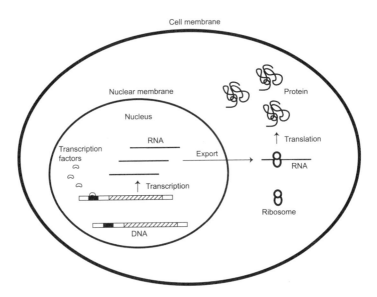

8.3. Steps in protein synthesis. A portion of DNA encoding one gene and its regulatory regions is depicted. The transcribed region is striped, whereas the regulatory switch is filled.

Like DNA, RNA molecules are chains of nucleotides, each containing a sugar, a phosphate group, and a base. RNA incorporates the same bases as DNA, except that uracil replaces thymine. Despite these similarities, DNA and RNA differ in functionally significant ways. DNA molecules resist fragmentation because they consist of paired strands of nucleotides containing the unreactive sugar deoxyribose. Stable double-stranded DNA makes an excellent archive for the blueprints that contain instructions for making proteins. RNA, in contrast, consists of single strands of nucleotides containing the more reactive sugar ribose. Though single-stranded RNA degrades more easily than DNA, its responsive structure makes possible a wider range of functional roles. In this chapter, we focus on messenger RNA (mRNA), RNA molecules that carry information from DNA to the ribosomes that synthesize proteins. Chapter 11 describes some of RNA's other functions.[15]

Suppose a kidney cell wants to make more EPO protein? First, the DNA strands surrounding the EPO gene separate and one strand acts as a template for the synthesis of mRNA.[16] Each mRNA molecule represents a copy of the EPO gene, containing the blueprints to make EPO protein. These messenger RNAs leave the nucleus and travel to ribosomes, which translate them into the amino acid backbones of proteins (Figure 8.3). Many mRNAs can be transcribed from the same DNA and many proteins can be translated from the same mRNA, so turning on transcription can quickly increase EPO protein levels. On the other hand, because mRNA quickly degrades, EPO protein synthesis declines soon after kidney cells turn off transcription of the EPO gene.

What determines EPO's transcription rate? DNA contains not only sequences that code for proteins, but also regulatory sequences. The evolutionary biologist Sean Carroll aptly named these control regions "switches."[17] These switches are flipped by transcription factors—proteins that bind to DNA switches, regulating transcription of nearby genes (Figure 8.3). For example, when hypoxia-inducible factor (HIF) attaches to a switch near the EPO gene, EPO transcription activates.

An elegant feedback pathway regulates EPO synthesis according to oxygen levels. Kidney cells constantly produce HIF, potentially generating a persistent signal for EPO production. Without some control, overproduction of red cells would result. But kidney cells also synthesize the enzyme hydroxylase, which requires oxygen to function. Hydroxylase adds a chemical group to HIF, targeting it to proteasomes—cellular garbage-disposal units that chew up proteins. When red cells deliver adequate amounts of oxygen to the kidneys, hydroxylase consigns HIF to proteasomes. As a result, EPO levels remain low, preventing unnecessary red-cell production. But when oxygen levels fall, the inhibition of red-cell production by hydroxylase weakens. For example, serious bleeding depresses red-cell content, decreasing the blood's capacity to carry oxygen. The resultant decline in cellular oxygen inactivates hydroxylase, freeing HIF to stimulate EPO transcription and thereby restoring red-cells levels.[18]

Training at altitude is a permissible alternative to illicit doping. The thinner air at altitude contains less oxygen, causing hypoxia and triggering red-cell synthesis through the HIF-EPO pathway. But there's a snag: low-oxygen levels impair training. Athletes can't exercise as hard at altitude, so the benefits of expanded red-cell volume may be countered by reduced training intensity. A tempting strategy is to "live high and train low," reaping the benefits of hypoxia by living at altitude, but returning toward sea level to maximize training efforts.

Athletes can live high and train low by residing in the mountains and traveling to the lowlands to train. Unfortunately, this approach suffers serious drawbacks. Most folks can't relocate, and even at the best locales, athletes spend most of each day traveling between high and low altitudes. Hypoxic chambers—living spaces that simulate altitude—offer an alternative. An athlete can sleep in a hypoxic chamber, then step outside it and exercise at sea level. Studies using hypoxic chambers paint a complicated picture: most show benefits, but some don't. These discrepancies may be partly due to the wide variability in individual responses to altitude. Some athletes, possibly owing to their genetic makeup, gain little from the live high–train low approach—they are nonresponders. Also, living in a small hypoxic chamber constrains lifestyle in ways that may be unacceptable to many athletes, especially over the long term. Once again, a one-size-fits-all recommendation seems unlikely to emerge. Athletes need to make individual decisions about altitude exposure based on their goals, training history, and, perhaps someday, their genetics.[19]

Aerobic metabolism seizes about 50–60 percent of glucose's energy by forging ATP molecules. This captured energy powers muscle contraction and other cell functions.[20] The remaining portion of glucose's energy dissipates as heat energy that can't be put to use. Some of this inefficiency is inevitable—whenever energy converts from one form to another, some useable energy transforms to unusable heat. But surprisingly, mitochondria sometimes activate mechanisms that lower metabolic efficiency, favoring heat production at the expense of ATP synthesis. These pathways contribute to divergence in metabolic rate among individuals, underpin adjustments to cold climates, and influence how exercise affects body composition.

Variation in metabolic efficiency is possible because the mechanism of ATP synthesis is indirect. Recall that electron carriers transfer high-energy electrons from the citric acid cycle to molecules on the mitochondrial inner membrane. These molecules strip energy from the electrons, using the energy to pump protons into the space between the inner and outer mitochondrial membranes (Figure 8.1). This proton pumping establishes a strong gradient that favors proton flow back across the inner membrane. Protons can cross the membrane through ATP synthase proteins, making ATP. Or they can travel through uncoupling proteins (UCP), generating heat without synthesizing ATP.

ATP synthase works like the pumps we described in Chapter 6, except in reverse. Rather than using ATP to establish differences in ion levels across a membrane, the synthase employs the proton gradient to synthesize ATP. Two regions form the ATP synthase. One portion—embedded in the inner mitochondrial membrane— spins when protons flow through it. This rotation twirls a second domain—residing in the mitochondrial interior—powering synthesis of ATP from adenosine diphosphate (ADP) and phosphate.[21] Approximately ten protons pass through the ATP synthase to make three ATPs. In summary, the ATP synthase is a molecular motor, its rotation coupled to the synthesis of ATP.[22]

Mitochondria convert the chemical energy of nutrients to an electrochemical gradient, then to mechanical energy—rotation of the ATP synthase—and finally back to the chemical energy of ATP. At each step, some useful energy dissipates as heat, partly explaining aerobic metabolism's imperfect efficiency. Inefficiency also arises from the activity of uncoupling proteins, which provide an alternate pathway for protons to cross the mitochondrial inner membrane, a detour around the ATP synthase (Figure 8.1). Compare uncoupling proteins to a car's clutch. Active uncoupling proteins allow mitochondria to burn nutrients without synthesizing ATP, much as your car's engine burns gasoline without driving the wheels when you press on the pedal and disengage the clutch.

Why would cells take the trouble to metabolize nutrients all the way through aerobic metabolism and then at the last moment short-circuit ATP synthesis? Or to put it another way, why do cells have uncoupling proteins? Uncoupling proteins contribute to temperature regulation because metabolic reactions still produce heat even when they bypass the ATP synthase. The thermoregulatory value of uncoupling proteins was first recognized in hibernating animals. Brown fat—specialized fat deposits rich in uncoupling proteins—packs the bodies of many hibernators. In contrast to metabolically inert white fat that expands human waistlines, brown fat actively burns energy, producing lots of heat but not much ATP—just what hibernators need to prevent dangerous drops in body temperature during periods of inactivity. Human infants also have plenty of brown fat, helping them counter the rapid heat loss that results from their large surface areas relative to their masses. And researchers recently identified brown fat deposits in adults. Activity of adult brown fat increases during cold exposure, consistent with a role in body temperature regulation.[23]

Know someone who can pack away food and lay around on the couch without gaining weight? Your annoying friend may have a wasteful, inefficient metabolism resulting from high uncoupling protein activity in brown fat. Uncoupling accounts for up to 30 percent of our total metabolism, so variations in uncoupling protein amount and activity considerably affect energy balance.[24] Unsurprisingly, brown fat correlates with body composition—people with high levels tend to have lower body mass indices. Brown fat also declines as we age, possibly contributing to our tendency to pack on pounds later in life.

Uncoupling proteins dwell not only in brown fat, but in most tissues, including skeletal muscle. The function of these proteins remains controversial. They may offset the damaging effects of oxygen, transport fatty acids, regulate glucose metabolism, or control the rate of ATP synthesis. And, like the uncoupling proteins in brown fat, these proteins may help explain why some people find weight control easier than others.

Variation in the UCP3 gene, which codes for an uncoupling protein produced in skeletal muscle, correlates highly with body weight. In the UCP3-55T allele, thymine replaces cytosine in a switch region of the UCP gene. Transcription factors bind the mutated switch differently, leading to increased synthesis of the UCP3 gene. So this mutation affects how much UCP3 skeletal muscles produce, rather than altering the structure or function of the protein that the UCP3 gene encodes. Because UCP3 levels increase, carriers of the mutation readily burn energy during sleep, lose fat easily during training, and have low body-mass indices. Human populations that live in cold environments express UCP3-55T at high frequencies, consistent with a role for uncoupling proteins in heat production. In temperate climates, people with the mutated allele might get away with an extra scoop of ice cream because their muscles are tiny furnaces, constantly burning nutrients.[25]

Though a wasteful metabolism might sound like a blessing, athletes should be cautious about wishing for one. The opposite condition—low activity of uncoupling proteins caused by a thrifty, high-efficiency metabolism—might be an advantage when resources are scarce, for example, during starvation or exhaustive exercise. Thrifty metabolisms maximize the rate of ATP synthesis, squeezing as much ATP as possible from each glucose molecule. So athletes might benefit from thriftiness, especially in their muscle cells. Studies support this notion: endurance training results in decreased UCP3 production, shifting muscles toward a thrifty metabolism. Although this thriftiness may improve exercise performance, tamping down the muscle's metabolic furnace might also counter the weight-loss benefits of exercise.[26]

However, recent research offers a final twist to the UCP story: although exercise decreases UCP3 in skeletal muscle, it has the opposite effect on UCP1—the uncoupling protein made by brown fat. Exercise stimulates release of irisin, a newly discovered hormone. In mice, irisin "browns" white fat, shifting lethargic white fat into energetic brown fat. Irisin's mechanism isn't fully understood, but it probably works by activating transcription of UCP1. By increasing brown fat deposits, irisin may underpin some of exercise's healthful effects, including its well-documented capacity to counter metabolic disorders like obesity and diabetes. Certainly, irisin has intriguing potential as a drug, but much more work will need to be done first. In the meantime, you can trigger irisin release—and deposit some beneficial brown fat—by heading out to the gym.[27]

Regulation of both the activity and the amount of proteins controls aerobic metabolism. Modulating existing proteins—as when molecules bind and morph hemoglobin—results in quick but short-term changes in protein activity. For example, recall from Chapter 5 the rapid modulation of phosphofructokinase and pyruvate dehydrogenase. On the other hand, changes in protein amount—as when transcription of uncoupling proteins increases or HIF protein degrades—take effect more slowly but have enduring consequences. Both types of regulation matter to athletes. Acute changes in protein activity drive us forward when the gun goes off, whereas changes in protein amount govern the training responses that prepare us for race day.

Chapter 9

Run Like a Woman

For the past three miles, I have been running with a small group of men. Back at mile 20, we were laughing, chatting, and supporting each other. Now we're grim, just barely hanging on, too tired to talk. We might finish the marathon, we might even still be running when we cross the line, but it's not going to be pretty. Yet despite our distress, we're still passing runners—mostly other men who went out too fast and are now shuffling.

And then the women start gliding past us. Each one offers a smile, words of encouragement, and the chance—sometimes even a friendly challenge—to match her pace. Believe me, we try. For a few strides, maybe even a block or two, we stick with her. Soon after she pulls away, another woman passes, and the cycle repeats. I pledge—again—to run my next marathon like a woman, to finish with style and strength.

I've heard similar stories from other male runners. Do they reflect real differences between the way men and women run marathons, or do men just tend to notice when women pass us? A quick analysis of split times in two recent marathons supports the anecdotes: as a group, men run faster than women, but women slow down less in the second half of marathons than men.[1] More is at stake than bruised male egos. Over the past two decades, exercise scientists have documented many physiological differences between women and men, with implications for training, nutrition, and race strategy.

Research that explores differences between human groups sometimes creates controversy. One source of uneasiness arises because dividing people along demographic lines tends to mask variation within groups. For example, some men accelerate at the end of marathons; some women crash. So at some point in this chapter, you may find yourself thinking, "That may describe most women (or men), but it's not true of me." Fair enough. When I talk about women and men in this chapter, I mean the average woman or most men, not all women or every man. Nor do I intend to imply that differences between groups necessarily constrain our potential. I am happy to report that—with attention to my training and race strategy—I have run marathons with negative splits, running the second half faster than the first. Current work on sex differences in physiology aims to learn more about both sexes by comparing their differences, with the intent to improve health, athletic performance, and medical treatment. I offer this chapter about sex differences in metabolism in the same spirit.[2]

Many athletes consume energy gels—ready-to-eat doses of sugary calories—during distance events. Most gels contain a mix of easily digestible carbohydrates, including simple sugars such as fructose and short sugar chains such as maltodextrin, a molecule with three to twenty linked glucoses. Gels offer an apparently uncomplicated way to improve athletic performance: just rip open the convenient package and squeeze the gooey gel into your mouth. Digestive enzymes break maltodextrin into its constituent glucose molecules, which enter your body across the intestinal wall, raising blood-glucose concentration. But the fate of this glucose—and whether the gel will deliver its promised boost—depends on when you ingest it.

Both at rest and during vigorous exercise, our bodies tightly regulate blood glucose. Later in the chapter, we examine what happens during exercise; let's start by looking at the resting condition. Suppose you eat a gel packet as a midmorning snack—perhaps unpalatable and nutritionally empty, but a simple example for us to explore. Blood-glucose levels rise, triggering release of the hormone insulin by the pancreas. Insulin activates responses in muscle, fat, and liver cells that store the extra blood glucose. Picture insulin as a diligent housekeeper, clearing the clutter from the living room and stowing it into a closet's storage bins. First, insulin sweeps extraneous glucose out of the blood into cells. Inside cells, insulin tucks glucose away by promoting its synthesis into compact, well-organized energy-storage molecules. Thus insulin controls a negative feedback loop: increased blood glucose leads to glucose storage inside cells, returning blood glucose to normal.

Cells absorb glucose through glucose transporters—membrane proteins that usher glucose across cell membranes. Glucose transporters share properties with the ion pumps and channels we discussed in Chapter 6. Like pumps, they work by a bind-flip-release mechanism, grabbing glucose on one side of the membrane and releasing it on the other side. However, pumps split ATP and use its energy to move ions either into or out of cells, sometimes against electrochemical gradients. In contrast, glucose transporters do not use ATP and so only allow net movement of glucose down its concentration gradient—either into or out of cells. Hence, the total rate of glucose transport depends on its concentrations on both sides of the membrane. Glucose flows faster into cells when its level rises outside (for example, after a meal) or when level falls inside (for example, when muscles burn glucose during exercise.) In this way, glucose transporters resemble ion channels, which also move molecules along, rather than against, their gradients.

Insulin controls glucose transport by regulating where glucose transporters reside, a mechanism that we have not yet encountered. To carry glucose into a cell, glucose transporters must be on the cell's outer membrane. But in the absence of insulin, most transporters inhabit small membrane spheres called vesicles (Chapter 7) located just under the membrane (Figure 9.1). These vesicles resemble those that contain neurotransmitters, except that transporters embed into the vesicles' membranes, whereas neurotransmitters float free inside the vesicles in neuron synapses. If the outer cell membrane resembles a battlefield—the boundary where the cell's interior confronts its exterior—then the transporters in the submembrane vesicles are like reinforcements— well-trained, fully equipped troops ready to join the fray. Insulin

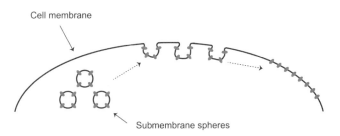

9.1. Activation of glucose transport. Glucose transporters
are depicted as gray ovals.

provokes the vesicles to fuse with the outer membrane, delivering glucose transporters to the cell surface. This mechanism activates transport much faster than making new proteins from scratch, because it quickly repositions presynthesized glucose transporters onto the outer membrane where they can function.[3]

All cells import glucose for their own use. Muscle, liver, and fat cells also store some of the glucose they soak up. These cells confront similar choices to a child given a quarter. The child can immediately spend the coin on candy, or put the coin into the bank where it might combine with other quarters to make a dollar. Similarly, cells can spend glucose by burning it to make ATP, or they can save glucose by tucking it away as glycogen or fat (Figure 9.2). Regardless of glucose's fate, the first step is the same: the enzyme hexokinase attaches a phosphate to glucose, forming glucose phosphate.

Muscle cells decide whether to burn or store glucose phosphate based on the availability of ATP and glucose. For example, when athletes begin to exercise, ATP levels in their muscle often diminish. Decreased ATP activates the enzymes involved in glycolysis, burning glucose phosphate to replenish ATP (Chapter 5). In contrast, during recovery from exercise, muscle cell ATP levels return to normal and blood-glucose concentrations often rise, especially after we eat a meal (or a gel packet.) Under these conditions, the enzyme glycogen synthase—aptly named because it synthesizes glycogen—catalyzes the joining of glucose phosphate molecules into glycogen.

Two pathways activate glycogen synthase, representing both types of regulation (allosteric and covalent) that we discussed in Chapter 5. First, when glucose supply exceeds demand, glucose level increases inside cells, leading to more glucose phosphate.

Glucose phosphate directly activates glycogen synthase by binding to a regulatory site on the enzyme, an example of allosteric regulation. In this way, glycogen synthesis increases when glucose phosphate levels rise inside cells. Second, rising blood glucose elevates insulin, which binds to cell membrane receptors. Activated insulin receptors stimulate a phosphatase that removes a covalently attached phosphate group from glycogen synthase, morphing it into an active form. As a result, muscles store glucose in the form of glycogen when blood-glucose and insulin levels rise.[4] Because of regulation by both allosteric and covalent pathways, glycogen synthase responds to concentrations of glucose both inside cells and in the blood.

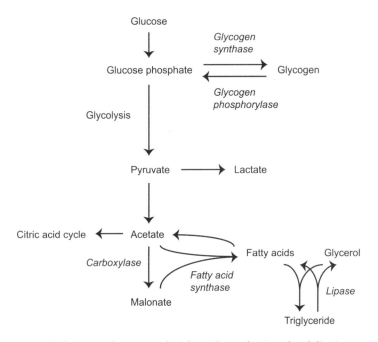

9.2. Pathways in glycogen and triglyceride synthesis and mobilization. Enzymes are in italics.

Glycogen chains pack into granules that contain 20,000—50,000 glucose molecules radiating out from a central protein called glycogenin. These compact granules stuff many glucose molecules into a small space. The majority of glycogen granules reside in muscles, both between and within sarcomeres, where they can quickly provide fuel to contracting cells. Well-rested muscle cells contain enough glycogen to power about two hours of exercise. The liver stores lesser—but still substantial—amounts of glycogen.[5] Both liver and muscle also store energy as fat, as anyone who has enjoyed pâté or bacon knows. Specialized fat storage cells—adipocytes—stockpile even more fat. Therefore, when we need energy during exercise, our bodies choose among energy reserves stored in different tissues and in different forms.

Glycogen readily breaks into glucose molecules, which are a flexible and powerful muscle fuel. Because glucose can drive either lactate fermentation or aerobic metabolism, muscles can use it across all exercise intensities. Moreover, we require glucose to perform maximal exercise: we can reach our peak exercise intensity by burning glucose alone and cannot attain top power output without it. In contrast, fats suffer several disadvantages compared to glycogen. Some tissues, including the brain, cannot use fats to make ATP. Also, glycolysis followed by lactate fermentation requires carbohydrate fuel, so fats cannot be used to produce ATP in the absence of oxygen. Finally, fats cannot deliver ATP as quickly as glycogen for two reasons: (1) the complex machinery needed to use fats has limited capacity and (2) burning fat requires more oxygen than burning glycogen. Consequently, we can reach only 60–70 percent of our maximal exercise intensity using fat alone.[6]

Unfortunately, we store only limited glucose in our muscle and liver glycogen, and risk running out of it during extended exercise.

On the other hand, our fat reserves contain a nearly bottomless supply of dense fat droplets. Even skinny people amass enough fat energy—in fat, liver, and muscle cells—to power days of moderate exercise. Fat stores energy efficiently because it is much denser than glycogen. Glycogen readily binds water, so each gram (g) of glycogen contains more than 2.5 g of water. Fat molecules, which are hydrophobic, exclude water and therefore form exceptionally compact energy packages.

Picture your muscle glycogen tank with its gauge pointing toward full at the start of a long race. Hours later, as the tank nears empty, you "bonk." You may suddenly hit the wall or slowly crash and burn. Perhaps blood-glucose levels drop, causing disorientation by limiting your brain's ability to make ATP. Or maybe your muscles begin to rely mostly on fat, reducing their power. Regardless of specifics, an empty glycogen tank means suffering and failure. Without glycogen, aerobic metabolism is less efficient and lactate fermentation is impossible. Can gels delay bonking? It certainly seems as if they might, either by providing an external glucose supply during exercise or by topping off the glycogen tank when taken before or after exercise.

I'm walking from my hotel to the starting line, five gels bouncing in my fanny pack. Though the marathon starts in forty-five minutes, my limbs already tingle with nervous energy and my heart rate prematurely accelerates. I imagine my glycogen supplies dwindling, wasted on my prerace anxiety and the trek to the start. Though my plan had been to eat a gel every five miles, I begin wondering whether I should swallow one now. At the very least, ingesting a gel would lighten the pack that was already chafing my back. And a little extra sugar couldn't hurt, right?

Actually, it could. Although eating carbohydrate in the hour preceding exercise temporarily elevates blood glucose, it also causes glucose levels to plunge below normal once exercise starts. Two mechanisms combine to increase glucose uptake into cells so much that blood-glucose levels fall below normal. First, insulin rises in response to the initial increase in glucose, fostering movement of glucose transporters to the cell membrane. Second, during exercise, signals within contracting muscle cells—including the rise in calcium that triggers muscle contractions and the fall in ATP that accompanies exercise—also translocate glucose transporters to the cell surface.[7] As a result, more glucose flows into cells, lowering blood glucose. Surprisingly, low blood glucose does not impair the performance of most athletes, though some athletes do suffer when glucose levels drop. Thus athletes should be cautious about their diet just before exercise.[8]

Athletes should consider both meal timing and composition. Eating a small carbohydrate-rich meal (equal to about 2–3 gels) three hours before exercise enhances athletic performance, probably by elevating liver glycogen levels. Insulin and blood glucose return to normal three hours after a meal, reducing the risk of fluctuations in these molecules that could counter the meal's benefits. But even hours before an event, many athletes need to avoid eating hard-to-digest foods. Foods with high fiber or protein content can produce gastrointestinal distress, which can impair performance.

Although preexercise meals require caution, the period immediately after exercise offers a risk-free opportunity to boost energy stockpiles. When endurance exercise depletes muscles of glycogen, they resemble a sponge squeezed dry of water. During the two hours following exercise, muscles aggressively soak up glucose and synthesize glycogen. Athletes who eat carbohydrate (equal to 1–2 gels) immediately after exercise synthesize far more glycogen than those who eat the same amount two to four hours after exercise. The composition of postworkout meals matters too: eating a bit of protein along with carbohydrate helps to accelerate glycogen synthesis.[9]

Although eating a gel forty-five minutes before a marathon could cause trouble, research supports my original plan—eating gels during the race. Glucose supplementation during exercise helps, especially in events lasting at least two hours. Ingested glucose replaces some of the glucose that would otherwise come from liver and muscle glycogen supplies, preserving glycogen and thereby postponing fatigue. Intriguingly, some research shows benefits from simply rinsing the mouth with glucose solutions. Perhaps tasting glucose reassures the brain that fuel is available to power exercise? Little evidence points to sex differences in responses to sugar ingestion during exercise; most studies show that women benefit, apparently by mechanisms similar to those of men.[10]

Time of day also influences the effectiveness of meals. We've all heard the advice to consume a good breakfast, avoid snacks before bed, and eat small meals throughout the day. Scientists have discovered connections between circadian rhythms and metabolism, supporting the notion that meal timing makes a difference. A clock that ticks in our hypothalamus—a region of the brain that controls reproduction, body temperature, and metabolism—strongly influences our eating patterns. Feedback among a set of proteins causes their levels to cycle with an approximately twenty-four-hour period (hence the term "circadian," from the Latin *circa* meaning "about" and *dies* meaning "day"). Light sets the circadian clock, locking it onto a twenty-four-hour rhythm, but the clock also interacts intimately with food intake and metabolism. Irregular mealtimes alter the clock's rhythm, whereas signals from the clock affect how we process the food we eat.

Levels of many metabolic enzymes vary throughout the day. In rodents, mRNA encoding glycogen synthase peaks at the beginning of the dark cycle—the start of the active period for nocturnal rodents, roughly equivalent to our morning. Differences in glycogen synthase and other enzymes mean that animals handle food differently at different times of the day. For example, rodents that eat high-fat meals during the daylight hours—when they are normally inactive—gain more weight even when they consume the same amount as animals eating during the night. Humans experience similar effects, after taking into account that our day is the rodent night. Thus athletes should pay attention to what time of day they eat meals. You might handle a big breakfast differently at 4:00 a.m. on race morning than at 10:00 a.m. before a training run, so testing your body's response to an early breakfast might be a good idea.[11]

Eating a normal diet only fills our glycogen tanks about halfway. To top the tank, athletes need more than a few well-timed gels before or after exercise. Hence the pasta dinner—a prerace tradition at many marathons—where competitors gorge on carbohydrates. Yet even a big plate of pasta fails to maximize glycogen storage. Instead, athletes must regulate the quantity of carbohydrates they eat over several days. Moreover, sex differences exist, so men and women may need somewhat different strategies.

Filling the glycogen tank requires reductions in training along with huge intake of carbohydrate. Young men can effectively carbo-load by eating 75 percent of their calories as carbohydrate for three or four days. For a 155-pound athlete, that's more than four pounds of cooked rice per day![12] Eating 75 percent carbohydrate requires rigorous removal of additional protein and fat from meals because carbohydrate-rich foods contain other nutrients. For example, pasta is about 80 percent carbohydrate, so carbo-loading competitors might need to forgo the pile of parmesan on their spaghetti. Such eating can be stressful for both athletes and those around them. Just ask my family. After carbo-loading for a day or so, I get jittery—probably because of extra dietary sugar—and I get cranky—maybe because the fat in my normal diet boosts my mood. But the benefits of carbo-loading before events lasting longer than ninety minutes may justify some unpleasantness. Carbo-loading postpones fatigue, enabling 20–45 percent increases in exercise duration. In events of fixed distance such as 5k runs, times improve about 2–3 percent when competitors start with filled carbohydrate tanks.[13]

For years, scientists, coaches, and athletes assumed that women could carbo-load the same way as men. But when physiologists began to test carbo-loading in women, the studies told a different story. Women increase their muscle glycogen and performance only slightly or not at all when they eat 75–80 percent of calories as carbohydrate for a few days. Why don't women show the same benefits as men?[14]

One possibility is that women don't ingest enough carbohydrates when they simply increase the percentage of carbohydrates in their diet. Perhaps carbo-loading guidelines need to emphasize the *amount* of carbohydrates, rather than the *percentage?* Studies support this idea. Women can increase their glycogen stockpile by eating a high-carbohydrate diet *and* increasing their total energy intake. Yet even when they do increase glycogen levels, women may not enjoy the same performance benefits as men. So compared to men, women may find it more difficult, and perhaps less valuable, to expand glycogen stores.[15]

Can we conclude that it takes more carbohydrate for women to carbo-load than it does for men? The question is tricky to answer because of a vexing issue that plagues comparisons between the sexes: the difficulty of accounting for differences in body weight. The problem is that women are not only lighter than men, but also have a higher body-fat percentage. When scientists normalize quantities such as VO_{2max} and food intake based on total body mass, men's values tend to be higher. But this finding doesn't mean that men are necessarily more fit or more voracious than women. Discrepancies between the sexes typically disappear when scientists normalize values by fat-free mass, probably because fat neither uses much energy nor produces any work during exercise. Though more work needs to be done, the current research shows that even when carbohydrate intake is normalized to fat-free mass, women need to eat more to fill their carbohydrate tanks, suggesting a real sex difference.[16]

Most guidelines recommend limiting carbohydrate loading to a few days before an event. One reason is that liver cells can convert excess dietary carbohydrates to fat. I learned this the hard way by eating a high-carbohydrate diet through an entire training season, always trying to keep my glycogen stores high. Maybe I did, but the diet had a frustrating side effect: I gained weight. Rather than watching pounds melt away as my mileage increased, I found my middle expanding. Growing love handles on a high-carbohydrate diet is possible for two reasons: (1) ingesting extra carbohydrates limits the need to burn fat, and (2) some excess carbohydrate converts to fat, especially after glycogen reserves have been fully loaded.[17]

To transform sugar into fat, liver cells first metabolize glucose to pyruvate by glycolysis and then convert pyruvate to acetate through the activity of pyruvate dehydrogenase, mirroring the steps of aerobic metabolism (Figure 9.2). In aerobic metabolism, the citric acid cycle dismantles acetate. Instead, fatty-acid synthesis builds acetate into fatty acids.[18] After an enzyme called carboxylase catalyzes the first step of the conversion of acetate into fatty acids, a huge enzyme called fatty-acid synthase finishes the job. Fatty-acid synthase contains two large (270 kDa) subunits, each with numerous distinct regions, including six different active sites.[19] Fatty-acid synthase is thus a conglomerate—numerous individual enzymes linked together into a single "superprotein" that catalyzes the last six reactions of fatty-acid synthesis.[20] Compare this pathway to glycolysis, where ten different proteins catalyze ten reactions. One advantage of grouping an entire pathway on the same protein: products efficiently pass from one reaction to the next because each step is adjacent. Grouping reactions together also simplifies and coordinates their regulation. If the amount of a superprotein increases or if it morphs to an active form, the whole pathway accelerates.

Fatty acids can combine with a molecule of glycerol to form energy-dense triglycerides, which are packed into storage sites in muscle, liver, and fat cells. Women store fat more readily than men, in part because of increased sensitivity to insulin, which promotes fat accumulation. Insulin stimulates carboxylase and fatty-acid synthase, the two key enzymes that catalyze conversion of carbohydrate to fat. Insulin also blocks lipolysis, the breakdown of triglycerides to glycerol and fatty acids, decreasing the use of fat as metabolic fuel and necessitating increased use of carbohydrate. Both mechanisms result in less glucose available for synthesis into glycogen, so differences in insulin sensitivity might partly explain sex differences in carbohydrate loading. In addition to insulin, sex hormones—discussed later in the chapter—probably also contribute to women's tendency to amass fat.[21]

Excess fat can impair performance in some sports. Consider the up-and-down motion of running. We lift our body weight more than a thousand times each mile, so each pound matters. On the other hand, good health requires adequate fat stockpiles. We need fats to build the lipids that form cell membranes as well as to synthesize certain hormones and other biomolecules. Fats are also an energy reserve, insurance against times of scarcity. Unsurprisingly, we have mechanisms that monitor and regulate fat stores.

Fat cells release the hormone leptin, which serves as an indicator of the body's energy reserves. Leptin binds to receptors in the hypothalamus, which orchestrates secretion of hormones controlling reproduction and metabolism. Falling leptin levels, a signal of reduced fat depots, stimulate increased energy intake by enhancing hunger. Lowered leptin also decreases energy use, partly by downregulating the activity of mitochondrial uncoupling proteins (Chapter 8). And decreased leptin puts the brakes on reproduction by impairing secretion of sex hormones, thereby reducing fertility. Forgoing reproduction is another way to save energy—ask any parent about the energy demands of caring for offspring.[22]

Hang around with a group of athletes, and you'll find that the talk eventually turns to food. Athletes pay special attention to diet, aware that their choices affect performance. Although attention to proper nutrition mostly serves athletes well, pressure to maintain a certain body size or shape can put athletes—both male and female—at risk of disordered eating. Female athletes in individual endurance sports (such as running) and judged sports (such as gymnastics) are particularly susceptible to chronic negative energy balance. Consequences can include impaired athletic performance, disrupted menstrual cycles, osteoporosis, and other serious medical problems.[23]

For much of the twentieth century, many marathons did not permit women to participate. The Boston Marathon first allowed women to compete in 1972; women ran the Olympic marathon for the first time in 1984. Pam Reed's success at the Badwater Ultramarathon—a 135-mile race across Death Valley, perhaps the hardest footrace on earth—highlights the absurdity of barring women from endurance events. In 2002, she beat the entire field, besting the nearest competitor by more than four hours. Even the most open-minded male athletes dislike losing to women, so you can imagine that men resolved to beat her in 2003. But Pam Reed won the race again, joined in the top five finishers by two other women. And it's not only superwomen like Pam Reed who excel at distance; when men and women are matched according to marathon finish times, the women tend to outperform the men at ultramarathon distances. The main sex difference we've discussed—the increased tendency for women to store fat—does not explain this success of women in ultramarathons. No one runs out of fat energy during a race, even in a 135-mile ultramarathon. On the other hand, sex differences in how athletes *use* fat during exercise might be an explanation.[24]

Our body fluids contain only two to three teaspoons of sugar, enough to power just a few minutes of activity, so we must mobilize stored fuel during prolonged exercise. Muscles seeking fuel to burn have options, in the same way that you have choices when you cook a meal. Just as you might raid your cupboards for ingredients, muscles can use their own glycogen and fat stores. Alternatively, just as you might head out to the deli or supermarket for supplies, muscles can get energy from liver glycogen stores or from the almost limitless deposits of fatty acid in adipose tissue. Muscles excel at matching the fuel to the situation in the same way that a thrifty chef selects ingredients carefully, dumping economical canola oil into the deep fryer while saving costly olive oil to drizzle on the bread. Our muscles favor fats—which are unable to sustain the highest exercise intensities—during low-intensity and long-duration exercise, whereas they prefer glycogen—which is in limited supply—during exercise of high intensity but shorter duration.

During exercise, female athletes rely more on stored fat than their male counterparts do, a possible advantage in distance events. By burning more fat, women limit glucose use, preserving glycogen stores and postponing fatigue. To return to the kitchen analogy: by liberally using canola oil, women stretch the olive oil a bit farther. The benefit of using fats explains the response of men to endurance training; they become more like women—at least in terms of fat metabolism—shifting their fuel use toward fat. Increased reliance on fat by women might be partly a consequence of their expanded fat stores. For example, studies show that women have more intramuscular fat than men and use it in proportion to its starting level. But women not only have more fat than men, they also mobilize their stored fat more readily during exercise than men. In short, women store fat more readily than men during rest, then burn it more readily during exercise.[25]

The hormone adrenalin stimulates the "flight-or-fight" response, triggering changes that prepare the body for action. Adrenalin shuts down fat synthesis by inhibiting carboxylase and mobilizes glycogen and fat from stores into the bloodstream by activating both glycogen phosphorylase and lipase. During intense exercise, adrenalin increases more in men than in women. Although this hormonal difference seems like a possible explanation for the elevated fat use in women, lower adrenalin in women should decrease, rather than increase, mobilization of fat by women.

Sex hormones—estrogen and progesterone in women and testosterone in men—are obvious candidates for mediating sex differences. But scientists struggle to unravel which hormones underpin sex differences because multiple hormones differ between men and women. Scientists can exploit natural variation in hormone levels, such as those related to aging or the female menstrual cycle. For example, estrogen rises in the mid-luteal phase (which precedes menstruation) and falls in the mid-follicular phase (which follows menstruation), so if estrogen influences fat metabolism, we would expect differences across the monthly cycle. Though some work points toward increased fat metabolism when women exercise during the mid-luteal phase, other work shows no effect of the monthly cycle. Scientists can also manipulate hormone levels, though they need to do this with care in human subjects. In one study, men supplemented with estrogen burned more fat during exercise compared to controls, suggesting that estrogen does contribute to differences in fat metabolism.[26]

Estrogen exerts its actions via a different mechanism than other hormones we've discussed. The sex hormones are steroids—derivatives of cholesterol, a much-maligned but indispensable fat that also contributes to the structure of cell membranes. Because steroids are fat-based, they pass readily through cell membranes. Most steroid receptors, including estrogen receptors, reside inside cells rather than on cell membranes. Estrogen receptors act as transcription factors after they bind estrogen, influencing synthesis of mRNAs encoding particular proteins, including those involved in fat metabolism. Levels of enzymes that catalyze fat metabolism differ between men and women, between stages of the menstrual cycle, and between men with and without estrogen supplementation. Thus regulation of gene expression by estrogen partly accounts for sex differences in fat metabolism. Unsurprisingly, testosterone also plays a role—when testosterone levels increase, fat accumulation decreases.[27]

At an aid station somewhere near mile 20 of the Cincinnati Flying Pig Marathon, a volunteer handed me a raspberry-flavored gel and I ate it. The day was hot, my training was inadequate, there were hills ahead, and I was fading. The gel, according to the science we've discussed in this chapter, should have helped. Instead, its cloying sweetness stuck in my mouth, kicking off waves of nausea. I walked, stumbled, and weaved my way across the bridge from Kentucky to Ohio and missed my prerace goal by ninety seconds. My cardinal mistake was that I ate a raspberry gel for the first time (and honestly, probably the last time) during that race.

Studies that take into account sex, age, and other demographic factors have helped fine-tune our understanding of human physiology. But even the most convincing evidence about your demographic bracket cannot unerringly predict how you will respond to a particular diet. Try everything yourself before the event or be prepared for devastating consequences: disappointing finish times, and if you are particularly unlucky, lasting aversions to perfectly innocuous fruity flavors.

Chapter 10

Drinking Games

Minutes before my first marathon, my wife, Kathy, sized up the competition and accurately noted, "Everyone here is skinnier than you." Although unhelpful, this observation wasn't as discouraging as her comment hours later as I passed the 18-mile post. The summer day had become muggy, and I must have been showing the effects. Kathy sent me out for the remaining eight miles with the admirably honest remark, "You always struggle in the heat."

For the record, despite these early missteps, Kathy has been remarkably supportive of my running. And in all fairness, nothing that she could have said would have helped on that June morning. The blame lies squarely on my shoulders. In graduate school, I spent years immersed in the science of fluid balance, so I should have known how to drink during exercise. But that morning, I comprehensively botched my fluid intake. After the race, my errors were easy to identify: too little water, too little sugar, and too little salt. Nonetheless, many years passed before I developed a strategy for effective hydration during long runs on hot days.

What makes proper drinking difficult? At first glance, the best strategy seems simple: just replace what we lose. During exercise, we lose water across the lungs as we breathe; we lose salt and water when sweat glands perspire and the kidney makes urine; and we lose glucose when muscles burn˙ it. During most exercise, our only way to restore these lost materials is by ingesting drinks. Therefore, the composition of exercise beverages affects how effectively we replace lost salt, water, and sugar.

I'm tempted to pitch this chapter as a quest for the perfect sports beverage, the ultimate exercise elixir, the ideal mix of ingredients for replenishing exercising bodies. But as with diet and training, no single all-purpose drink recipe guarantees success. Environmental conditions, the type and duration of exercise, and variation among individuals all influence the amount of salt, sugar, and water athletes lose. Moreover, the uptake of drinks into the body also depends of many factors. The properties of epithelia—cell layers that separate the internal and external environments—determine how quickly molecules move into and out of the body, thereby influencing which drink compositions and hydration strategies are most effective.

As a beginning marathoner, I suffered in the heat partly because I had a drinking problem. I often hurried though aid stations, grabbing a cup and splashing its contents across my face so that sticky liquid ran down my chin, wetting my shirt, and in the most embarrassing instances, soaking my shorts too. Obviously, beverages cannot work their rehydrating magic when you're wearing them. Less obviously, successful ingestion is only the first step in a long journey from the cup to the bloodstream. Fluid sloshing in the stomach aids performance no more than fluid dripping from the nose, because the gut's inside—its lumen (cavity)—is physiologically outside the body (Figure 10.1). If you find this hard to swallow, picture the gut as a tube extending through the body; anything that remains in the gut stays outside the body. To enter the body, molecules must cross a cell layer called the intestinal epithelium.

Epithelial layers regulate transport into and out of our bodies much as stone walls once controlled movements into and out of medieval cities. Mortar holds stone walls together, dividing the stones into inward- and outward-facing surfaces. Likewise, structures called tight junctions attach epithelial cells to each other, separating each cell's membrane into two domains: one facing outside the body and the other facing inside. Hydrophobic lipids pack the core of cell membranes, repelling water. Therefore, to cross the intestinal epithelium, water must either slip between cells or flow across both inward- and outward-facing cell membranes through membrane proteins called water channels. Depending

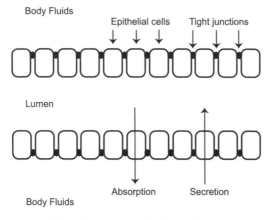

10.1. Diagram of epithelial-cell layers.

on the abundance of water channels and the tightness of junctions between cells, epithelial layers can be either tightly sealed against water movement or quite leaky to water.

Even in leaky epithelia, water does not flow unless pushed by some force. In Chapter 6, we discussed the sodium pump, which harnesses adenosine triphosphate's (ATP's) energy to move sodium ions out of cells, even when sodium's electrochemical gradient favors inward movement. No analogous protein pumps water actively across cell membranes. Rather, water can either be pushed by fluid pressure in the way that blood pressure propels the circulation of blood (Chapter 1) or driven by differences in osmolarity—the concentration of dissolved solutes. If you stir a teaspoon of sugar into your tea, the sugar dissolves and increases the tea's osmolarity. Dissolved solutes attract water, so water flows toward solutions with high osmolarity. Moreover, water does not play favorites among different solutes, moving just as readily toward glucose, for example, as it does toward sodium. In other words, the concentration of dissolved particles, not their makeup, determines water movement along gradients in osmolarity. Osmotic gradients drive fluid movements across epithelial-cell layers in the intestine, kidneys, and sweat glands.

Whether beverages cross the intestinal epithelium and nourish your body or remain in your gut and slosh uncomfortably depends partly upon their osmolarity. Beverage osmolarities vary widely. Osmolarities of water that flows in rivers or out of municipal taps range from 1 to 10 milliosmolar (mOsm). Most sports beverage osmolarities hover near those of our body fluids, between 280 and 310 mOsm. Fruit drinks and sodas are more concentrated than body fluids, with osmolarities ranging from 500 to over 1,000 mOsm. We might expect plain water to be absorbed more quickly than drinks with higher osmolarity because the large osmotic gradient between water and our body fluids favors absorption. Conversely, the gradient for absorption dwindles as osmolarity of fluids in the gut lumen rises toward osmolarity of body fluids, and reverses when gut fluid osmolarity exceeds that of body fluids.[1]

In practice, a complicated relationship exists between drink osmolarities and water absorption. One reason is that the osmolarity of gut fluids does not exactly match that of the most recently ingested beverage. Our mouths secrete saliva and our guts secrete other salty solutions, adding water and solute to ingested fluids. Also, gut enzymes digest nutrients into smaller components. For example, a drink containing 10 grams (g) of maltodextrin—short chains of glucose—has lower osmolarity than a drink containing 10 g of glucose because the linked glucoses in maltodextrin count as a single solute molecule. However, enzymes in saliva and intestinal secretions quickly digest maltodextrin into individual glucose molecules, so its effect on intestinal osmolarity is nearly the same as the drink containing glucose. For this reason, a drink's total amount of easily digestible nutrients—in other words, its energy content—strongly influences intestinal absorption.

A final source of complexity is that intestinal water absorption partitions into two steps: an early passive step and a later active one. Drink composition affects these steps differently. Sugars—the most abundant solutes in most sports beverages—impair the early step and stimulate the later one. To untangle this complex interaction between water absorption and sugar concentration, let's compare the absorption of water to that of sugary drinks.

Ingested water cascades down the esophagus into the stomach, where it awaits delivery to the intestine. The stomach stores fluid, metering it out to the intestine in manageable portions through the pyloric sphincter—a ring of smooth muscle between the stomach and intestine. By the time ingested water passes through the pyloric sphincter, it has picked up some salt from salivary and gastric secretions. Nonetheless, its osmolarity remains below that of the body fluids, resulting in an osmotic gradient for water absorption. Water passively flows along this osmotic gradient because the intestine has a leaky epithelial layer. Thus the early portion of the intestine rapidly absorbs fluid after we drink water. In contrast, drinks with high osmolarities, such as sodas and fruit juices, oppose this first step in fluid absorption, possibly even causing fluid to flow into the intestinal lumen.

When the initial section of the intestine stretches or when it contains fluids of high osmolarity, it releases hormones that dampen stomach contractions and tighten the pyloric sphincter. Thus drinks rich in sugar—especially those with higher than 8 percent sugar—tend to slow gastric emptying rate, the pace of fluid transfer from stomach to intestine. Gastric emptying caps the amount of fluid that we can comfortably drink, limiting our capacity to hydrate. Drinking faster than the maximal gastric emptying rate of about 1.2 liters (L) per hour simply distends the stomach, causing gastric distress rather than hydration. So a trade-off exists in sports beverage composition: drinks with sugar deliver needed energy to exercising muscles, but too much sugar delays gastric emptying, limiting the amount of beverage that we can comfortably ingest.[2]

Let's return to the fate of ingested water. As fluid flows farther down the intestine, its osmolarity progressively rises because passive water absorption leaves solute behind. Consequently, intestinal osmolarity increases toward body fluid osmolarity, decreasing the osmotic gradient and thus slowing passive absorption. Eventually, active absorption replaces passive absorption. To absorb water actively, the intestinal epithelium carries solutes from the intestinal lumen into the body, generating an osmotic gradient. In other words, the intestine hauls solute and water follows along. We see this principle in action throughout the chapter; sweat glands and kidneys also propel water by transporting solutes.

When active intestinal absorption goes awry, severe consequences can ensue. Every day, about 9–10 L of water enter our guts. We eat and drink 2 to 3 L of fluid, whereas salivary, gastric, and intestinal secretions contribute another 7 L of salty fluids. When all is well, our guts absorb almost all of this fluid. On the other hand, disorders such as cholera that impair gut function can rapidly produce life-threatening dehydration. Sufferers of cholera can lose up to 2 L of fluid each hour because of accelerated secretion of salt and water into the gut and reduced absorption of materials out of the gut. Although cholera annually kills 100,000 to 120,000 people, oral rehydration solutions—beverages designed to counter severe dehydration—save many lives. These rehydration beverages contain sugar, which greatly improves their effectiveness by stimulating active absorption. Although the stakes may be lower, sports drinks apply the same principle. To see why sugars enhance active water absorption, we need to peer into the intestine's mechanism for transporting solute.[3]

Intestinal epithelial cells couple sodium transport to movements of glucose and other sugars. Let's first examine how this coupling works in a simple spherical cell with a sodium pump on its membrane. Sodium pumps carry sodium out of the cell in exchange for potassium, expending the chemical energy of ATP to generate a large electrochemical gradient for sodium to enter the cell. Suppose the cell's membrane also contains sodium-glucose cotransporters. These membrane proteins resemble glucose transporters (Chapter 9), except that they tether sodium and glucose, necessitating that they move together across cell membranes.

Sodium-glucose cotransporters resemble a taxi poised to carry a married couple to the airport. Just as neither partner will leave without the other, sodium will not flow through sodium-glucose cotransporters unless glucose shares the ride. And just as one spouse will typically pay the full cab fare, a strong sodium gradient can pull glucose against a concentration gradient when a cotransporter couples their movement—sodium pays the fare for glucose. So although sodium-glucose cotransporters do not directly expend energy, they can accumulate glucose inside cells by exploiting the sodium gradient created by the sodium pump. If simple glucose transporters also exist on the cell membrane, they will carry glucose in the other direction—down its concentration gradient and out of the cell. In the hypothetical cell just described, glucose cycles futilely, entering cells through cotransporters and leaving through transporters (Figure 10.2, above).

On the other hand, if we place the same proteins on the correct sides of intestinal epithelial cells, they will pump sodium and glucose from the gut lumen toward the body fluids. If you enjoy puzzles, try positioning each protein on inward or outward facing epithelial-cell membranes to create a solute-absorbing epithelium. The solution is as follows: sodium pumps and glucose transporters on the inward-facing membrane, sodium-glucose cotransporters on the outward-facing membrane (Figure 10.2, below). In this configuration, the sodium pump drives sodium toward the body fluids, generating a gradient for sodium to flow from the gut lumen into the epithelial cells. Sodium then partners with glucose to cross the outward-facing membrane of epithelial cells through cotransporters. If glucose abounds in the intestinal lumen, sodium and glucose uptake accelerates, causing glucose levels to rise inside cells. Glucose then flows through transporters out of cells across the inward-facing membrane, joining the sodium moved that direction by the sodium pump. Consequently, solutes accumulate inside the body, establishing an osmotic gradient for water absorption. Without glucose in the intestinal lumen, the whole process ceases. Hence, beverages containing sugar stimulate the solute movement that underpins active water absorption.[4]

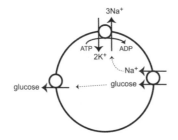

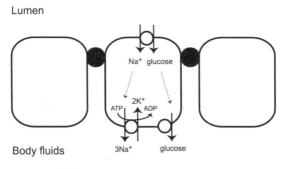

10.2. Transport of glucose. Above: Single cell. Below: Epithelial cell layer.

The bottom line is that drinks with too much sugar slow gastric emptying and impair passive water absorption because they create an osmotic gradient opposing absorption. But the intestine also absorbs drinks without sugar slowly because active water absorption follows coupled movements of glucose and sodium. Thus the Goldilocks principle applies: the best sports drinks contain just the right amount of sugar. Drinks with 6–8 percent sugar hit the sweet spot, delivering energy to exercising muscles and accelerating active water absorption while only minimally slowing gastric emptying.

Active water absorption requires sodium as well as glucose, so do rehydration drinks need sodium too? It depends. Gut secretions contain sodium, so additional sodium may not be needed to stimulate water absorption. And because most of us carry lots of extra sodium, replacing lost sodium typically is unnecessary during short exercise bouts. On the other hand, sodium levels largely determine osmolarity of our extracellular fluids, the fluids outside our cells. Extracellular fluids include the fluid that bathes our cells and the plasma, the watery portion of our blood outside red cells. Sodium levels also affect the volume of our extracellular fluid by influencing thirst and kidney function. For these reasons, inclusion of sodium in exercise beverages improves their effectiveness during exercise exceeding one or two hours, especially in hot conditions.

While training for my first marathon, I often envisioned a triumphant finish, my own little Olympic victory. I understood that the last miles would be tough, but I imagined myself crossing the line with style. My dreams, I believed, were realistically modest. I wasn't looking to finish with a showboating sprint, only with enough pace to demonstrate that I had battled adversity and emerged victorious. Unfortunately, my actual finish was far more modest than my dreams. Not long after Kathy reminded me of my sensitivity to heat, my running stride shortened to a jog, then to a shuffle, and then nearly to a crawl. I probably mustered a slightly enhanced shuffle at the finish, but I fooled no one, especially not myself. The hot day had demolished me.

Heat challenges athletes for numerous reasons. Exercising muscles produce heat, increasing the body's heat production by up to ten-fold and threatening to raise body temperature. High body temperatures alter protein structure and damage cell membranes, impairing functions of almost all organs, including the brain. Unsurprisingly, we have mechanisms to regulate body temperature, but unfortunately this regulation comes at a cost. When outside temperatures are below body temperature, we dissipate extra heat simply by sending blood toward the skin, where it dumps heat to the environment. Rosy cheeks help keep us cool, but increasing skin blood flow strains the circulatory system's capacity to send enough blood to exercising muscle. And as more blood flows elsewhere, less blood may return to the heart, impairing the heart's capacity to increase its output (see Chapter 1). When we exercise in the heat, heart rate rises more than it does during exercise at neutral temperatures because the heart labors to meet demands for blood flow to both skin and muscles.[5]

When outside temperatures rise above body temperature, sweating becomes the only means to dissipate heat. Sweat glands dump heat by secreting fluid onto the skin. When this fluid evaporates, the transition from liquid to gas consumes heat energy, cooling the blood flowing near the skin. Although a powerful thermoregulatory mechanism, sweating causes at least two problems. First, sweating dehydrates us, reducing blood volume and further exacerbating the problems of blood return to the heart and flow to the muscles and skin. Second, sweating perturbs the osmolarity of body fluids. Sweat is less salty than blood, so sweating initially raises blood osmolarity.

Though other animals also dissipate heat by evaporating body fluids, humans lead the pack. Dogs and many other mammals pant, evaporating water across only a small surface area—their tongues and upper respiratory tracts. And effective panting involves short breaths that hamper gas exchange, a particular problem when exercising muscles demand more oxygen. Thus sweating dumps heat better than panting, especially during exercise. From a thermoregulatory perspective, humans differ from other mammals in both obvious and subtle ways. Most noticeably, we lack thick hair over most of our body. Although this absence of insulation helps us dissipate heat, it also exposes our skin to the direct rays of the sun. Less conspicuously, eccrine sweat glands cover our skin, whereas apocrine sweat glands dominate in most other mammals.

Apocrine sweat glands collaborate with sebaceous glands to produce oily sweat that smells when acted upon by bacteria. Horses—one of the few mammals other than humans that can run long distances in heat—produce a thick frothy sweat from these glands. Although apocrine glands do cool horses and other hairy mammals, their sweat output is lower than eccrine glands, limiting their thermoregulatory effectiveness. When humans become nervous, overactive apocrine glands in our armpits and groin secrete stinky sweat, but these glands contribute little to our thermoregulation. Instead, our sweat flows mostly from eccrine sweat glands, which secrete a thin watery sweat that evaporates easily, thereby readily dissipating heat. Our prodigious sweat flows from about two to four million of these eccrine glands distributed over most of our thinly haired bodies. Chimpanzees and other primates, in comparison, have more restricted distribution of eccrine glands—mainly on their hands and feet—and thus must rely on panting.[6]

Copious sweating enables humans to exercise for long durations in the heat. Consider, at the extreme, the Badwater Ultramarathon, a 135-mile run across Death Valley in the July heat. Don't ask your German shepherd for company when you attempt Badwater, and don't worry about being beaten by a cheetah. Without adequate sweating capacity, other mammals can't take the heat. Thus our abundant sweat glands join the elongated Achilles tendon (Chapter 1) as features that are sometimes maligned, yet distinctly human. In fact, scientist Nina Jablonski has argued that "plain old unglamorous sweat has made humans what they are today."[7]

Sweat glands—infoldings of the skin lined with a layer of epithelial cells—lovingly fashion sweat in a two-step process. The first step resembles active fluid absorption by the intestine, with two differences: fluid moves out of the body rather than into it, and the glands transport salt (sodium chloride) rather than sodium and glucose. At the base of sweat glands, cells actively transport salt into a coiled tubule. In this region, water passes easily across the leaky epithelial layers, following salt into the gland. The resultant precursor sweat matches the osmolarity of blood because water flows into the gland until osmolarity of the glandular fluid equalizes with the blood. In the second step, precursor sweat flows out of the gland through a narrow duct. Epithelial cells lining this duct reabsorb salt, diluting the sweat. The gland's duct is less leaky to water, so water does not follow salt and the sweat remains dilute. Both steps—the initial secretion of salt and its subsequent reabsorption—require energy because sweat glands pump salts against their electrochemical gradients. We pay a cost both to make sweat and to dilute it, another downside of sweating.

Suppose you rehydrate with a drink that exactly matches the salt concentration of your sweat? If you drink an amount equal to your sweat volume, both fluids and salts would be replaced at the rate they are lost. A path to perfect rehydration? Unfortunately, although this strategy seems scientifically rigorous, it suffers several shortcomings. First, during hard exercise in the heat, athletes can sweat about 2 L per hour, exceeding typical gastric emptying rates. At such high sweating rates, drinking cannot keep pace with sweating. Second, completely replacing lost salt and water is unnecessary because small changes in body fluid volume or osmolarity have little effect on performance. Dehydration, for example, is most problematic when losses exceed 2 percent of body water. Unnecessary drinking wastes precious time. Third, matching the salt concentration of sweat produces unpalatably salty drinks. Manufacturers try to cover some of the saltiness of sports drinks with flavor, but athletes may drink less when drinks become too salty. Sweat sodium levels range from 20 to 120 millimoles per liter (mmol/L), lower than the blood levels of 140 to 145 mmol/L. Sports drinks typically contain no more than 20 mmol/L sodium, less than all but the most dilute sweat.[8]

A final reason why drinks that exactly match sweat are impractical is that sweat content not only varies markedly among individuals but also within individuals under different conditions. For example, sweating rate influences sweat composition.

When we sweat rapidly, fluid moves quickly through the sweat gland duct, leaving little time for salt reabsorption. As a result, sweat salt levels rise when sweating rate increases. Two or three weeks of workouts in the heat also influences sweating, increasing sweating rate by about 15 percent, but lowering sweat sodium levels by about 50 percent. Thus, even though sweating rate increases with training, reabsorption of sodium from sweat increases more, and total sodium loss decreases. Our sweat glands, which already tend to conserve salt, cling to salt even more avidly after heat acclimation, even though making dilute sweat tends to increase blood osmolarity. In sum, exercising humans hoard salt.[9]

Before a big-city marathon, runners follow well-established rituals. The day before the race may include a trip to the expo, a pasta dinner, and a restless night in a hotel bed. Race morning may consist of an early arousal to drink and eat, a slow walk to the starting line, and a long wait for a Porta Potty. Consider all the precious salt and water—which the runners will soon desperately need—that ends up staying behind in these portable latrines when the gun goes off. But if racers ingest a bit of salt along with their water, urine production may slow, resulting in smaller contributions to the toilets and better hydration. Additionally, salt stimulates thirst, encouraging fluid ingestion. Thus, while exactly matching the salt concentration of sweat is unnecessary, inclusion of some salt in sports beverages improves their effectiveness, especially during exercise lasting for many hours.

Salty drinks promote fluid retention partly through the actions of vasopressin—a hormone released by neurons into the bloodstream.[10] Vasopressin acts on blood vessels and kidneys, influencing several interrelated variables: body-fluid volume, blood pressure, and extracellular fluid osmolarity. In this chapter, we focus on regulation of osmolarity, which is crucial for two reasons. First, altered osmolarity triggers changes in cell volume, and both cell shrinkage and swelling can have devastating consequences. For example, shrunken epithelial cells may be an ineffective barrier between the body and the environment, whereas swollen cells may press on adjacent structures, harming their function. Second, abnormal osmolarity caused by altered sodium levels perturbs electrical activity in cells such as neurons and muscle because electrical signaling depends on sodium gradients between cells and their surrounding fluids.

Because sweat contains proportionally more water than salt, plasma osmolarity rises during exercise. The higher solute concentration outside cells attracts water, causing water to leave. As a result, cells shrink. Although every cell is subject to such shrinkage, osmoreceptor neurons scattered throughout the brain actively respond, initiating a reflex pathway that controls plasma osmolarity. Stretch-sensitive ion channels—ones that open when tension along the membrane decreases—trigger this response. Membrane tension declines when cells shrink, so the stretch-sensitive channels in osmoreceptors open, allowing both sodium and potassium to flow. The resultant changes in membrane voltage activate vasopressin secretion. Osmoreceptor neurons located in a brain region called the hypothalamus synthesize vasopressin, transport it to the pituitary gland through long axon projections, and release it into the bloodstream. Another set of osmoreceptor neurons regulates thirst—activating an urge to drink when osmolarity rises. To recap, exercise in the heat elicits sweating, increasing osmolarity of body fluids. As a result, osmoreceptor neurons shrink, causing stretch-sensitive channels to open and triggering thirst and the release of vasopressin.[11]

Vasopressin controls osmolarity through its effects on kidney tubules—the million or so tiny tubes in each kidney that modify urine before excreting it (Figure 10.3). Blood pressure forces fluid into one end of these tubules. At the other end, each tubule drains into a big tube called the ureter and then into the bladder, so anything that remains in the kidney tubule eventually leaves the body. Extraordinary volumes of fluid flow into our kidney tubules. Although the kidneys comprise less than 1 percent of our body mass, they receive about 20 percent of the heart's blood flow. And about 7.5 L (almost 2 gallons) of fluid filters from blood into kidney tubules each hour. Clearly our kidneys must recover most of this "pre-urine." Otherwise, the Porta Pottys would overflow on race morning and, even worse, we would cross the starting line

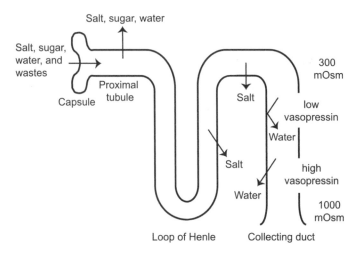

10.3. Diagram of a human kidney tubule.

already dehydrated. Vasopressin orchestrates water retention, ensuring that the kidney holds onto the right amount of fluid.[12]

To picture how the kidney works, imagine two different methods for decluttering your home. In the first, you carefully inspect each room, removing every unnecessary item. In the second, you empty the entire contents of your home onto your front yard and only recover what you want. (Optional extension of method two: advertise a yard sale and peddle your junk to neighbors.) Waste removal by the kidney resembles the second method. Blood pressure forces water, salt, sugar, and many other useful substances, but not proteins or red cells, into the kidney tubules (Figure 10.3). The first segment of kidney tubules receives this pre-urine and reabsorbs most of the valuable water and solutes. Cells much like those in the intestine and sweat glands transport salt, sugar, amino acids, and other solutes. Water follows the resultant osmotic gradient across a leaky epithelium. Wastes remain in the tubule, destined for excretion, like the rubbish that remains on your front lawn after you've repossessed everything useful.

The second part of the kidney tubule adjusts urine salt and water, thereby regulating both volume and osmolarity of body fluids. Contemplate this region's response to three different "diets": chips and pop, chips alone, or pop alone. If we ingest chips and pop, the kidney needs to excrete both salt and water. Our kidneys accomplish this easily, simply leaving salt and water in the tubule so that they are eventually excreted. If we drink pop alone, the kidney needs to dilute the urine by excreting fluid but retaining salt. Our kidneys dilute urine by the same mechanism that sweat glands dilute sweat; they reabsorb salt across epithelial layers that remain impermeable to water. Water stays in the tubule, and we produce copious, dilute urine. If we eat chips alone, our kidneys need to concentrate the urine by excreting salt but retaining water. Producing concentrated urine presents a challenge. We've seen how epithelial-cell layers can move water by first transporting salt. But concentrating urine requires that water be pulled out of the kidney tubule while salt remains.

Though all vertebrates have kidneys, only mammals make concentrated urine. The strategy is to draw fluid out of kidney tubules by surrounding them with salty fluid. Mammalian kidney tubules follow a tortuous path (Figure 10.3). After running along the kidney's surface, the tubule plummets down toward the kidney's center and climbs back up to the surface. This region—the loop of Henle—creates very high osmolarity in the kidney's interior. The last segment of the kidney tubule—the collecting duct—plunges back into this salty interior. Tubular fluid at the start of the collecting duct is dilute. As long as the duct remains impermeable to water, water stays in the tubule, resulting in copious dilute urine. When we need to conserve salt—for example, if we are on a "pop diet"—the collecting duct can reabsorb more salt, further diluting the urine.[13]

On the other hand, if the collecting duct's water permeability increases, the salty fluid deep in the kidney sucks water out of the tubule. When we need to conserve water—for example, during exercise in the heat—vasopressin triggers fusion of membrane vesicles containing water channels with the duct's epithelial-cell membranes, creating a path for water to flow out of the tubule.[14] The resultant fluid reabsorption concentrates the urine and decreases its volume, conserving the body's water. In summary, the later portions of kidney tubules separate salt and water movements in both space and time. The loop of Henle transports salt into the kidney's interior, creating an osmotic gradient that the

collecting duct either exploits to retain water or ignores to excrete water. In a separate process, the last portions of the tubule
either reabsorb salt or leave it in the tubule, adjusting the urine's
final salt content. Because the kidney regulates water and salt in-
dependently, it responds appropriately to the pop diet, the chips
diet, and everything in between.[15]

Our kidneys can limit perturbations, but not completely reverse them. To restore blood volume and osmolarity, we must
drink. Consider first what happens when we drink water during
exercise in the heat. As our intestines absorb water, body fluid
volume increases and osmolarity falls toward its starting level. At
first, our kidneys retain this fluid. But if we drink enough water,
osmolarity returns to normal, shutting off the signal to release
vasopressin and causing the kidneys to start excreting water. Now,
drinking more water produces no benefit because it lowers osmolarity further, causing vasopressin levels to drop even more and
resulting in even less water reabsorption by the kidney. Though
ingested water enters the body, it passes directly out the kidneys.
Even worse, because the kidney cannot make perfectly dilute urine,
some salt flows out with the excreted water, exacerbating the
problem.

Because the kidney carefully regulates blood osmolarity, our
extracellular fluid's salt content largely determines its volume.
Compare our body fluids to lemonade made from ready-mix
powder. The amount of properly sweetened lemonade you can
make depends on how much powder you have. Spill a bit of powder, and you need to either add less water to the remaining powder or tolerate diluted lemonade. Similarly, if we lose salt and water in sweat and replace only water, we either stop short of restoring
body fluids or dilute them. Because the kidney excretes water

when body fluids become dilute, failure to replace salt inevitably leads to dehydration. We see now why sweat glands conserve salt: holding onto salt helps us retain water.

During short-term exercise, drinking water may help restore osmolarity and bring volume toward normal. But as salt continues to leave in sweat, drinking water alone becomes progressively less effective. In fact, during exercise lasting more than a few hours, ingesting too much water can be life threatening. Excess drinking can cause hyponatremia—low blood sodium levels—a serious condition that has contributed to several deaths of marathon participants. Inappropriate vasopressin secretion probably exacerbates some cases of hyponatremia. If vasopressin remains elevated even when osmolarity falls, then the kidney continues to reabsorb water back into the body. Losses of salt in sweat, coupled with too enthusiastic drinking and improper retention of water, can lead to dangerous decreases in blood osmolarity. The most serious consequence is brain swelling caused by water movement from overly dilute body fluids toward the higher osmolarity inside brain cells. To prevent hyponatremia, athletes should limit fluid ingestion to reasonable levels. Again, the Goldilocks principle applies: appropriate drinking boosts performance; too much can impair it. To get this right, athletes may need to monitor their fluid ingestion carefully. Thirst is a notoriously poor indicator of hydration in humans, so athletes cannot rely on thirst to motivate their drinking.[16]

Ingesting some salt during extended exercise both reduces the risk of hyponatremia and improves rehydration. Replacing some of the salt lost in the sweat keeps blood osmolarity and sodium levels high, stimulating thirst and vasopressin release. Thus drinks with salt encourage continued water ingestion, minimize reductions in osmolarity and enhance water absorption. If some salt helps, will a boatload of salt help more? Why not chug some pickle juice the night before a big event? If you have the stomach for it, this strategy might be worth a try. Pickle juice powered the Philadelphia Eagles to a 41–14 victory over archrival Dallas on the opening game of the 2000 season, played in 109°F Dallas heat. Presumably, the salty brine helped the Eagles stay hydrated, improving their performance. Recent work shows that pickle juice may also offset cramps, but it empties slowly from the stomach, so athletes should probably use it with caution during exercise.[17]

Perhaps a swig of beer sounds more palatable than a shot of pickle juice? Unfortunately, alcohol blocks both the secretion of vasopressin and its actions on the kidney tubule. As a result, alcohol impairs water retention, leading to dehydration. For this reason (and probably many others), a postevent celebratory beer may be a better idea than a preevent anticipatory one.[18]

More than a decade after my first marathon, Kathy met me at mile 65 of the Mohican 100 trail run and offered me a few slices of black olive pizza—just what I needed to face the upcoming nighttime miles. Truthfully, I was looking forward to sunset. Daytime temperatures had climbed above 80°F, baking the course's country roads and forest trails. Yet, despite fourteen hours exercising in the heat, I felt immeasurably better than at mile 24 of my first marathon.

The difference? Certainly I was better prepared physically for the Mohican 100 after a decade of distance running. Also, I braced myself for a warm race day by training in winter gear on hot summer afternoons, braving not only the heat but also the questioning looks—and even some smart aleck remarks—from properly dressed people. But most important, I talked to other runners, learned their tricks, and then experimented myself with different combinations of water, sports drink, salt, and food. In the end, my strategy took into account the science of fluid balance, the wisdom of my ultrarunning friends, and years of my own experiences. I finished the Mohican 100 that year because I knew what would work for me at my current fitness level, at the pace I planned to run, and under the conditions that we were facing.

In the end, the optimal composition of a drink depends on external conditions and the intensity and duration of exercise. Sugar stimulates water absorption by the intestine; salt promotes water reabsorption by the kidney. But too much sugar, salt, or water can be counterproductive. Finding exactly the right balance in every situation seems impossible. Fortunately, getting it perfect isn't necessary. If we ingest enough salt, sugar, and water without greatly overdoing it, our bodies will readily make the final adjustments.

Chapter 11

More Gain, Less Pain

Ten days before the marathon, I complete the final lap of my last hard workout. The sacrifices of the past nine months—early mornings instead of late nights, carrots sticks instead of carrot cake, track workouts instead of fun runs—have paid off. I'm ready to race. Or, more accurately, I'll be ready to race next week. Right now, I'm drained. My walk is a slow limp, and I doubt my sore legs could run three miles. When friends ask whether I'm OK, I respond with as much confidence as possible. But inwardly I struggle to control the self-doubt, even though experience tells me that my body will rebound, that my torn-down muscles will recover, that in the end I will be stronger.

At the peak of a training season, athletes may be tempted to take solace from Friedrich Nietzsche's famous maxim, "What does not kill me makes me stronger," or Benjamin Franklin's argument, "There are no gains, without pains."[1] Though neither author was referring to exercise, athletes can relate to the message that reaching goals requires suffering. But most athletes also recognize that it would be unwise to base a training program solely on these maxims. What almost kills you might weaken you, and pain might mean injury, not gain.

Training athletes, like mountain climbers, need to balance risk and reward. Carelessly flirt with the precipice—train too hard, too long, or too frequently—and you risk fatigue and injury. On the other hand, hang back in safe territory—keep the weight and reps constant, or run the same loop at the same pace—and you won't see improvement. The trick to successful training is to push up to, but not past, the breaking point.

Each sport has its own training season with distinct timing and rhythm, yet a common pattern can be discerned. Long before a featured event, athletes are often in a preparatory phase, maintaining their fitness but not aiming for improvement. Next is an extended period of normal training, where workout intensity progressively increases, resulting in performance gains. As the event nears, athletes peak by overreaching—a short period of training at very high intensity. Overreaching for too long inevitably causes injury or overtraining—physiological and psychological fatigue that impairs performance. So the trick to effective training is to overreach without overtraining.

Though training seasons typically build from easy to grueling, it is a mistake to envision training as a staircase, athletes going just a bit harder each day until they reach a peak. To build up to a hundred push-ups, a bad plan is to start with a single push-up and add one each day until you reach your goal. Instead, all good training programs incorporate cycles of easy and hard, of stress and rest, of activity and recovery. A training month might include three progressively difficult weeks followed by a light week. A training week might incorporate tough days alternating with easy ones. And, at the core of all training programs, a period of recovery follows each workout. Although it's tempting to focus solely on the workouts, successful training requires proper recovery too.

Training resembles other long-term responses to stress, such as acclimation to altitude. Everest climbers progressively expose themselves to altitude, with short visits to high camps followed by returns to base, a pattern reminiscent of the easy-hard cycles of exercise training. Going to altitude too quickly or for too long can cause altitude sickness, just as overly intense exercise can cause overtraining. Like altitude, the stress of exercise creates signals that trigger lasting responses. Exercise depletes energy reserves, damages cells, elevates muscle calcium, and alters physical conditions such as acidity, oxygen, and temperature. These signals activate regulatory pathways that stimulate enduring changes, mainly by altering the type and amount of proteins made by cells. Recovery is an essential ingredient of training because protein synthesis happens mainly after exercise ends. In short, exercise tears us down; recovery builds us up.

Each athlete responds differently to training, so programs must be tailored to individuals, a process that often seems more art than science, more trial and error than systematic strategy. But personalized training prescriptions begin to seem more feasible as scientists unravel the pathways that activate training responses. Perhaps programs could be designed to specifically stimulate those pathways that produce the desired training response? Maybe signaling molecules could be monitored to assess the effectiveness of workouts or to follow the progress of training? Such fine-tuning might allow athletes to maximize the benefits of workouts while avoiding unnecessary efforts that elevate risk of overtraining.

Years ago, I skipped an afternoon run and instead joined a low-key game of community softball. The next morning I ached from head to toe. Even my legs were sore, though they were in tip-top shape for distance running. I had rediscovered specificity—one of the key principles of training. Chest presses won't improve your 5k time. Cycling won't improve your biceps curls. And distance running certainly won't prepare you for a friendly game of softball.

Comparing resistance training—lifting weights and similar activities—to endurance training highlights the specificity of training. Resistance training leads to better communication between muscles and motor neurons; contractile strength increases because muscles activate more fully (Chapter 6). A more noticeable effect is that muscles get bigger, mainly because resistance training packs them with extra actin, myosin, and other proteins of the muscle sarcomere (Chapter 3). With extra contractile proteins, additional crossbridges can form, and strength improves. In contrast, endurance training enhances aerobic capacity by increasing muscle mitochondria (Chapter 8) and capillaries—the tiny blood vessels that supply muscles.

To a certain extent, the responses to resistance and to endurance training are mutually exclusive—one reason why training is specific. Muscles crammed with mitochondria have less space for sarcomeres and vice versa. Also, muscle types express different myosin heavy chain paralogs (Chapter 4). The myosin variant expressed in slow-twitch muscles uses adenosine triphosphate (ATP) slowly, favoring endurance. In contrast, the variant expressed in fast-twitch muscles burns ATP quickly, generating high power but promoting fatigue. Cells can express myosin that is either powerful or fatigue-resistant, but not both. So athletes can aim for maximal endurance or maximal power, but cannot achieve both simultaneously. Consequently, athletes often seek to balance endurance and power in a way that optimizes performance in their sport.

Resistance training pumps up muscles by two mechanisms. First, training stimulates synthesis of muscle proteins, resulting in hypertrophy—the expansion of existing muscle cells. Second, training activates muscle stem cells called satellite cells, resulting in hyperplasia—an increase in the number of muscle cells. During development, skeletal muscle cells arise from fusion of precursor cells, resulting in mature muscle cells with multiple nuclei. In adults, activated satellite cells can fuse with each other, producing new muscle cells. Satellite cells can also fuse with damaged muscle cells, contributing new nuclei and thereby repairing injured cells. Fusion of satellite cells to existing cells may also stimulate hypertrophy, though some scientists question this role of satellite cells.[2]

Many factors control muscle growth, including two proteins released by muscle cells—myostatin and insulin-like growth factor (IGF-1). IGF-1 presses on the gas pedal, accelerating muscle growth. Myostatin pushes on the brake, inhibiting growth. Despite their opposing roles, these factors act by similar mechanisms. Both proteins bind to cell surface receptors, triggering signaling pathways that regulate transcription factors (Chapter 8), and thereby altering gene expression. Though both proteins can serve as hormones, traveling through the blood to cause widespread effects, they often target nearby muscle cells, or even the cell that released them. In this way, signaling by myostatin and IGF-1 resembles a reminder tacked to the fridge rather than a blog post—the message is intended for the writer's close associates, not the entire world. The local actions of myostatin and IGF-1 contribute to another kind of training specificity—the restriction of benefit to the muscles used during a workout. Doing curls with just your left biceps will cause it to release less myostatin and more IGF-1, promoting its own growth but leaving other muscles—including the neglected right biceps—mainly unaffected.[3]

Mice that lack myostatin display the ultimate beach bodies. Strains with myostatin gene knockouts have hypertrophy, hyperplasia, and a shift toward fast-twitch cell types, causing two- to three-fold expansions of muscle mass with no increase in fat. But although mice without myostatin may look great strutting along the boardwalk, you might not want them on your tug-of-war team. The overmuscled mice are not stronger than their leaner cousins, partly because of impaired aerobic metabolism resulting from fewer mitochondria. Similarly, cattle lacking myostatin are impressively muscled but suffer increased muscle damage in response to exercise. So complete knockout of myostatin does not improve strength and may even be harmful.[4]

On the other hand, moderate decreases in myostatin may increase muscle strength as well as mass. Although complete knockout of myostatin fails to increase muscle strength, more subtle genetic changes can be beneficial. For example, in whippets, a dog breed used for racing, speed correlates with a mutation that truncates the myostatin protein, impairing its function. Whippets with one copy of the mutation—heterozygotes—tend to be the fastest racers. On the other hand, dogs that carry the mutation on both copies of their myostatin gene—homozygotes—are not as fast. Likewise, anecdotal evidence suggests that humans who are heterozygous for a mutation in myostatin have increased strength, but those who are homozygous are not stronger. Also, some research shows that depressed myostatin levels may contribute to the strength gains of resistance training.[5]

Myostatin's role in muscle growth makes blocking it an obvious therapeutic strategy, especially as treatment for muscle weakness disorders. A recent clinical trial using antimyostatin antibodies to reduce myostatin succeeded in expanding muscle mass, though strength did not measurably increase. Athletes and coaches may be tempted to apply the same strategies to enhance performance, but such approaches run afoul of antidoping regulations.[6]

Myostatin highlights the challenges of translating basic science into therapies and training recommendations. Simply aiming to increase or decrease a factor such as myostatin may be too simplistic because the optimal dosage may be intermediate. Also, because muscle function demands coordination among many systems—metabolic, contractile, regulatory—altering one factor is unlikely to replicate the training response. So let's turn to IGF-1, another key factor in resistance training.

On the surface, IGF-1 plays a simple role. The strain of strong contractions causes muscles to release IGF-1, which acts on nearby cells to repair damage and prepare muscles for the next bout of exercise. IGF-1 binds to both mature muscle cells and satellite cells, promoting hypertrophy and hyperplasia. In contrast to the effects of myostatin knockout, increased strength accompanies the resulting expansion of muscle mass. Elevating IGF-1 levels seems like a sensible objective of strength training. Perhaps athletes and coaches could monitor IGF-1 and construct workouts that maximize its levels? Unfortunately, this approach is trickier than it seems because the messenger ribonucleic acid (mRNA) of IGF-1 comes in three versions that code for proteins with slightly different structures and functions.[7] These IGF-1 variants arise from the structure of the IGF-1 gene.

Genes are divided into distinct sections, like the chapters of a book. Exons—stretches of deoxyribonucleic acid (DNA) that code for proteins—alternate with introns—stretches of "nonsense" DNA that do not code for protein. When enzymes in the cell nucleus copy a gene into RNA, the RNA initially contains both coding exons and noncoding introns. This pre-RNA resembles an odd book where intervening chapters of gibberish interpose between lucid chapters. To make such a book coherent, we would need to tear out the nonsense chapters and reattach the remaining chapters. Similarly, cells must splice pre-RNA, that is, remove the introns and zip the exons into an mRNA that contains uninterrupted instructions for making an entire protein.

The spliceosome—a large enzyme complex that contains both RNA and protein—clips pre-RNA and stitches the exons back together. The RNA components of spliceosomes contribute to catalysis. In other words, these RNA molecules act as enzymes. RNA molecules play a similar role in ribosomes, catalyzing the synthesis of new protein molecules. Thus, although some RNAs are messengers that carry instructions for making proteins, other RNA molecules are enzymes that accelerate and coordinate the splicing of pre-RNA into mRNA and the subsequent translation of mRNA into protein.

Proteins make effective enzymes because their intricate structures bind reactants and encourage their transition to products (Chapter 2). DNA cannot catalyze reactions because its invariant double helix lacks the necessary structural complexity. In contrast, single-stranded RNA folds into varied shapes by base pairing with itself, forming structures such as stems and hairpin loops (Figure 11.1). Though RNA molecules fail to match the remarkable structural diversity of proteins, they do fold into shapes that are complex enough to catalyze or regulate reactions.

When making mRNA, spliceosomes sometimes skip exons or replace one exon with an alternate (Figure 11.2). To see why this matters, imagine what would happen if we assemble the chapters of a book differently. Omit a chapter or swap the last chapter with a different version, and the book's meaning might change. Simi-

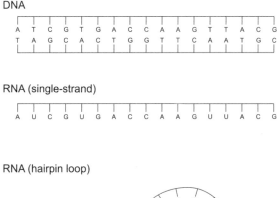

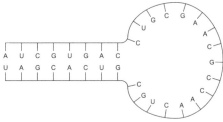

11.1. Ribonucleic acid (RNA) structure. Above: Section of double-stranded DNA. Middle: Section of single-stranded RNA. Below: Hairpin loop in an RNA strand.

larly, splice variants—the different mRNAs resulting from alternative splicing—often encode proteins with different functional properties. Most human genes have splice variants, greatly expanding the diversity of proteins the 20,000–25,000 genes in our genome produce.[8] Which variant a cell produces depends on many factors. Splice variants are often specific to certain tissues or developmental stages, and stressful conditions such as exercise may alter splicing.

Splice variants muddy the IGF-1 story. The six exons of the human IGF-1 gene are spliced into three variants: one missing exon 5, another missing exon 6, and a third variant with a partial

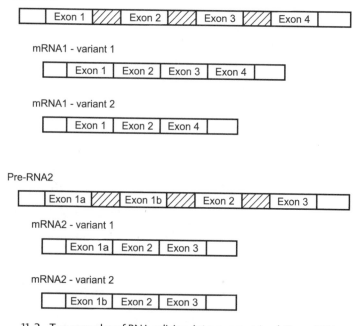

11.2. Two examples of RNA splicing. Introns are striped. Two mRNA variants for each pre-RNA are depicted.

fifth exon. The partial exon stops in the middle of a nucleotide triplet coding for an amino acid. When stitched to exon 6, this split triplet disrupts the gene's reading frame (Chapter 3), altering the sequence of the remainder of the protein. Some scientists call this splice variant mechano-growth factor because its level increases locally in exercised muscles, possibly triggering the specific training response. The other variants behave more like hormones, circulating through the blood and thereby activating training responses across the entire body. Controversy swirls around the roles of mechano-growth factor and other IGF-1 splice variants in the training response. My point in this instance is that practical applications of IGF-1 in training and medicine require a clearer picture of the roles of splice variants. The same holds true for many other proteins involved in exercise and training.[9]

Resistance training produces readily noticeable responses, such as ripped deltoid and pectoral muscles. In contrast, responses to endurance training are less visible. For example, no outward physical characteristic unites the runners who pour across the finish line of the New York City Marathon each year—they are short and tall, skinny and stocky, muscled and lean. Nonetheless, endurance athletes share distinct physiological traits. Big hearts, large blood volumes, and well-developed blood vessel networks power their cardiovascular systems. Meanwhile, aerobic enzymes and mitochondria power their muscles. Vascular endothelial growth factor (VEGF) triggers some of these endurance-training responses. As suggested by its name, VEGF promotes growth and differentiation of the endothelial cells that line blood vessels.

During exercise, the fall in oxygen levels, the stress in blood vessels resulting from increased flow, and other factors stimulate

VEGF production and secretion by skeletal muscle, blood vessels, and other tissues. VEGF acts locally, binding to cell surface receptors and triggering angiogenesis—proliferation of blood vessels—in the exercised muscles. Consequently, the network for delivering oxygen to these muscles expands, improving aerobic fitness. This role of VEGF in endurance training resembles the role of IGF-1 in resistance training. In both cases, factors released during exercise trigger training responses. And, just as splice variants tangle the IGF-1 story, splice variants of VEGF lend complexity to its functions.[10]

The VEGF gene contains eight exons, with the last three subject to alternative splicing. Although most of the splice variants diverge only slightly in function, alternate splicing of the eighth exon results in proteins that *inhibit* rather than promote angiogenesis. Several factors, including IGF-1, regulate splicing of VEGF. IGF-1 rises in exercising muscles, shifting VEGF splicing toward variants that activate angiogenesis. So exercise not only increases VEGF levels, but also may increase the ratio of angiogenic to nonangiogenic splice variants.[11]

The local effects of growth factors such as IGF-1 and VEGF partly explain why training primarily benefits the muscles that contract during workouts. Because exercising muscles release IGF-1 and VEGF that exert effects mainly on nearby cells, training responses are most vigorous in the muscles used during exercise. Specificity also arises because signaling pathways activate *within* exercising muscle cells. A protein called PGC1 (see note 12 for an explanation of this acronym) anchors these intracellular pathways during aerobic training, integrating signals and triggering responses.[12] Although PGC1 resides in many different cells, we focus on its role in muscle.

PGC1 orchestrates endurance training. During exercise, ATP falls and calcium rises inside muscles, stimulating PGC1 levels to increase (Chapter 5). PGC1 binds to and activates transcription factors that regulate genes involved in metabolism. When scientists overexpress these transcription factors using genetic tools, they produce ultraendurance mice. Compared to controls, these mice have more mitochondria, more slow-twitch fibers, and can run more than twice as far before fatiguing. By stimulating transcription factors, PGC1 conducts the endurance-training response.[13]

The story of PGC1's function is complicated by—surprise!—splice variants. Recent work shows that spliceosomes generate two PGC1 variants, PGC1a and PGC1b, with different first exons. At rest, PGC1a mRNA is about a hundred-fold more abundant than PGC1b mRNA. After two hours of leg exercise with restricted blood flow, PGC1a mRNA levels increase less than ten-fold, while PGC1b mRNA levels increase over a hundred-fold. In short, PGC1b responds more vigorously to exercise than PGC1a. Perhaps PGC1b contributes more than PGC1a to training responses? We need more information about the variants, especially analysis of how their functions differ, to assess such speculations.[14]

The effects of training are often slow to appear but long lasting. Running coach Robert Wallace compared long training runs to saving pennies: "Put them in a jar and at first it doesn't seem as if you have much. But after a few months, the volume begins to look impressive." A mechanism called DNA methylation could contribute to some of the enduring effects of training by altering gene expression. If regulation of gene expression by transcription factors resembles flipping a light switch, then DNA methylation is like pulling the plug. DNA methylation occurs when enzymes attach a methyl group—consisting of one carbon and three hydrogen atoms—to DNA, resulting in structural changes to the DNA that can silence the expression of nearby genes. For example, patients with diabetes display increased methylation of the PGC1 gene and express PGC1 mRNA at lower levels than nondiabetic controls.[15]

Recent research implicates DNA methylation in regulation of gene expression following exercise. In one study, six months of aerobic exercise altered methylation of more than 600 genes in fat cells. About a third of these genes were expressed at different levels following the training program. Another study found that a single bout of exercise decreased methylation and increased expression of several key genes in skeletal muscle, including PGC1. Scientists do not yet know how long these DNA modifications persist following exercise, so we need more research before concluding that methylation coordinates long-term training effects. Another area of uncertainty is whether the physical activity of prospective parents could influence the health and fitness of their future offspring. Maternal diet, cigarette smoking, and other nongenetic factors influence DNA methylation of growing fetuses, so it is plausible that exercise too could have an effect. And some methylation modifications may be inherited by offspring through egg and sperm, so exercise by parents might even affect offspring before their conception. Though these links between parents and children remain speculative, I'm grateful for every push-up and mile my parents logged.[16]

Imagine an exercise pill that would elicit the positive benefits of exercise without the need to leave the couch. Many athletes are uneasy with the notion. Athletes find training satisfying partly because they fight hard for the gains. And even though they may grumble about them, most athletes enjoy tough workouts, especially once they're completed. Taking an exercise pill seems a bit like cheating at solitaire; you might win, but so what? On the other hand, the public health benefits of an exercise pill would be astronomical because inactivity is a risk factor for many diseases.

We've seen that exercise triggers multiple regulatory pathways, that these pathways interact with each other, and that interventions that only influence a single pathway may not produce desirable outcomes. So an effective exercise pill would probably need to mirror many of exercise's effects—a daunting prospect. Up until very recently, we have been stuck studying one gene, mRNA, or protein at a time. Chapter 12 delves into the genome-wide approaches that are transforming exercise science by evaluating thousands of molecules at the same time. Let's preview that material now with an example related to this chapter's topic: new strategies that allow scientists to assess many mRNAs simultaneously.

Suppose scientists wished to determine which genes exercise activates in human muscles. Rather than study dozens of genes one at a time, they can isolate RNA from muscle before and after exercise, convert the RNA back into DNA (which is easier to work with than RNA), tag the DNA with a detectable marker, and then pass the labeled DNA across a chip that contains probes for thousands of genes. The abundance of each mRNA correlates with the amount of marker stuck to its probe on the chip.

A study published in 2010 compared levels of more than 20,000 mRNAs in men and women following resistance exercise. One finding is that before exercise, women express genes involved in fat metabolism at higher levels than men, consistent with other work that shows women burn fat more readily than men (Chapter

9). Another result is that after exercise, changes in gene expression in men last longer than they do in women. This difference may stem from a better capacity of women to restore homeostasis, thereby more quickly removing stimuli for altered gene expression. Whatever the cause, the more prolonged changes of gene expression in men may contribute to their larger gains in muscle mass after resistance training. Finally, although most genes showed the same pattern in women and men, genes involved in several signaling pathways displayed sex differences. In women, exercise activated genes involved in two signaling pathways known to inhibit muscle growth, one explanation for sex differences in responses to resistance training. For each of these conclusions, this study demonstrates differences in networks of genes, offering more powerful evidence than studies of single genes.[17]

Although evaluating every mRNA represents a huge advance, such approaches still do not paint a complete picture. So far, we have considered differences in abundance of mRNAs, RNA molecules that code for proteins. But other RNA molecules play regulatory roles. Small RNA molecules that do not code for proteins control many processes, including changes in mRNA levels in response to exercise. These regulatory microRNAs add another layer of complexity to the already elaborate systems that manage the training response. Whether the goal is to develop exercise pills or simply to understand training, microRNAs offer both promise and challenge.[18]

MicroRNAs—in association with the proteins slicer and dicer—stalk mRNA molecules, first targeting them for demolition and then cutting them to pieces. MicroRNAs are copied from DNA in the same way as coding mRNAs and then clipped into mature microRNAs of about twenty-two nucleotides by dicer and other enzymes. MicroRNAs bind to slicer, another RNA-cutting enzyme, forming RNA-protein complexes.[19] MicroRNA strands dangle from these complexes like lures, hooking mRNAs by pairing with their bases. The slicer protein shears snared mRNAs into oblivion, or in a less gruesome alternative, blocks their translation to protein. MicroRNAs regulate at least half of the genes in mammals, so they are a marquee feature of gene regulation rather than an obscure sideshow.[20]

Exercise and training alter microRNA levels. For example, a single bout of endurance exercise reduces the levels of a microRNA that represses PGC1 production. This removal of PGC1 suppression by microRNAs partly explains why PGC1 levels rise after exercise. MicroRNAs may also contribute to the differential effects of resistance versus endurance training. Endurance training increases expression of a microRNA that blocks IGF-1 synthesis, whereas resistance training decreases the same microRNA. Single exercise bouts and long-term training can also exert opposing effects on microRNAs—a study shows that acute exercise increases levels of certain microRNAs, whereas training decreases them. Finally, although microRNAs apply powerful effects in the cells where they are produced, recent work shows that they also leave cells and circulate through the bloodstream like hormones. Both exercise and training influence these circulating microRNAs, which may produce effects across the entire body.[21]

MicroRNAs also contribute to transitions between fast and slow muscle types. One remarkable mechanism involves sequences embedded in the introns of slow-twitch myosin genes. When spliceosomes remove these introns from the myosin pre-RNA, the short excised RNA sequences form microRNAs with hairpin structures. In mice, eliminating these microRNAs by genetic engineering results in more fast-twitch muscle cells, whereas overexpressing them converts muscles entirely to slow-twitch cells. The microRNAs from the slow-twitch myosin gene apparently activate synthesis of more slow-twitch myosin. The resultant rise in slow-twitch myosin then further increases levels of the microRNAs, leading to even more slow-twitch myosin. Such positive feedback could accelerate transition of muscle cells to the slow-twitch subtype.[22]

Differences in microRNAs levels might partly explain why some people greatly benefit from exercise programs, whereas others don't. A recent study evaluated microRNAs levels before and after a resistance training program. High-responders—people with large gains in muscle mass and strength—displayed different patterns of microRNA expression than low-responders—those who showed only small gains in strength and mass. So variations in microRNAs correlate with the amount of benefit people get from training programs. We need to interpret these results carefully. MicroRNAs might cause differences in the training response between people. Alternatively, the observed microRNA variations might be a secondary effect, a response to differences in the training response rather than a cause. Either way, such work recommends further exploration of microRNA's role in training.[23]

Scientists routinely produce microRNAs and use them to reduce production of specific proteins in animal studies. Similar strategies could be abused to improve human performance, especially when reductions in a protein—for example, myostatin—might result in improvements. Although considerable technical problems remain, scientists are developing medical treatments using microRNAs, so illicit uses in athletics are certainly plausible. Antidoping authorities have banned the use of microRNAs to improve performance, and the race is on to develop reliable tests to catch cheaters.[24]

Training is a psychological challenge as well as a physical one. During the overreaching phase, athletes must accept pain to achieve gains. But the phase that follows overreaching—taper— can be even more psychologically difficult. Considerable evidence documents that performance improves when athletes progressively reduce training in the weeks before an event. In general, the strategy is to decrease training volume but to maintain intensity. For example, runners, cyclists, and swimmers might decrease miles by 20–25 percent three weeks before a race, by 40–50 percent two weeks before the event, and by 65–70 percent in the week before racing. For athletes accustomed to pushing past prior limits, backing down this way can be hard to swallow.

Yet serious competitors need to cope with tapering because it can increase performance by 2–6 percent, enough to make the difference between the podium and the middle of the pack. Recent work illuminates the taper mechanism. Type IIa muscle cells—ones that are intermediate between slow and fast cells— expand during taper. In these cells, tapering alters levels of mRNA for genes that encode proteins involved in muscle growth. For example, muscle RING-finger protein-1, which triggers degradation of muscle proteins, decreases during taper, whereas muscle regulatory factor 4, which stimulates muscle protein synthesis, increases. As a result, muscle cross-sectional area expands and strength improves.[25]

Taper comes at a tough time for competitors. As the event nears, anxiety mounts, nerves fray, and self-doubt threatens—a cruel moment to be deprived of the release that a hard workout can bring. Perhaps athletes will rest more easily knowing that their muscle regulatory pathways are diligently building capacity above the levels that were achieved during overreaching. Gain without pain!

Chapter 12

Chasing the Holy Grail

Most athletes have a holy grail, a goal lofty enough to seem heroic but realistic enough to motivate us to keep trying. Some casual runners aspire to finish a marathon. Many serious recreational runners hope to one day qualify for the Boston Marathon. For the nearly elite, going to the Olympic trials might be the goal. For many years, my holy grail was a three-hour marathon.

Over a three-year span, I ran five marathons under three hours and twelve minutes but failed to break the three-hour barrier. Several times I passed the twenty-mile marker on track for a sub-three finish, only to bonk or cramp. Achieving my three-hour marathon goal began to feel unattainable. Then, on a crisp October morning, I finished the Columbus Marathon in two hours, fifty-nine minutes, and twenty-nine seconds. Why did I succeed on that day when I had previously failed repeatedly? There's no single answer. To run that time, I needed *everything* to go right: a year of training without injuries, a week of good health and restful sleep prior to the race, cool dry weather on race day, the right pace strategy, and the cheers of two friends at mile 23, who picked up my spirits when I was ready to quit.

Similarly, no single gene, protein, tissue, or organ determines our athletic success—many molecules cooperate to achieve even the simplest athletic moves. In this book, we've explored proteins individually, sometimes considering their interactions with a few other proteins. But many hundreds, perhaps even thousands, of different proteins collaborate during a hundred-meter dash, a home-run swing, or a reverse pike dive. Likewise, gene networks, rather than single self-sufficient genes, underpin our capacity to perform.

No longer satisfied with single-molecule studies, exercise physiologists increasingly seek approaches that monitor many molecules simultaneously. Recent technical breakthroughs make these tactics feasible. Such approaches point toward a new holy grail for exercise scientists: they now pursue a global synthesis of all the molecules involved in exercise.

Athletic ability varies remarkably among humans. Consider the finishing times at a typical 5k race. The winners might cross the line about sixteen minutes after the starting gun fires; many participants will finish between twenty-five and thirty-five minutes; and a few determined folks will take an hour or more to complete the race. Thus 5k times differ more than four-fold among the people who sign up to run road races. We'd expect even more variability if those disinclined to participate were included in the sample.

Why do 5k times vary so much? Demographic factors such as age and sex account for some variation. Behavioral factors such as diet and training also contribute. Some variation arises because multiple physical features (such as aerobic fitness, muscle strength, and running efficiency) as well as psychological factors (such as motivation and pain tolerance) influence finish times. But even when we measure well-defined physiological traits and account for obvious sources of variation—age, sex, motivation, training— big differences in athletic performance persist among humans. For example, aerobic capacity (VO_{2max}) varies widely among untrained adults, ranging from about 15 to 60 milliliter oxygen $\times$ kilogram body mass^{-1} $\times$ minute^{-1}.[1]

Scientists and athletes naturally wonder what portion of athletic ability comes from genes compared to environment—defined as all nongenetic factors, including diet, training, childhood experiences, and myriad other influences. To differentiate between genetics and environment, scientists typically compare relatives to people who are unrelated. Traits that are more similar in relatives than in nonrelatives have a strong genetic component. Unfortunately, relatives share not only genetic similarity, but also tend to experience similar environments, especially during childhood. For this reason, the most convincing studies contrast identical to fraternal twin pairs. Identical twins share the same level of environmental similarity as fraternal twins, but are genetically more similar than fraternal twins. Despite the technical challenges, studies consistently demonstrate that genetics affect athletic performance. The contribution of genetics varies from 20 percent to 80 percent depending on the subject population and performance measure. For example, genetics accounts for 40–60 percent of the variation in aerobic capacity of untrained people.[2]

Knowing how diet, training, and other environmental factors influence exercise has obvious practical benefits because we can manipulate these factors to maximize success. In contrast, we cannot (yet) easily change our genetics, so knowledge about the genetic basis of performance seems less useful. Our genetic endowments apparently constrain us, leaving us little recourse. But this dismal outlook only applies if we view genetics alone, ignoring crosstalk between genes and environment. Practical applications emerge when scientists account for interactions between genes and environment, such as how genetics influences athletes' responses to training, diet, altitude, and other environmental factors. Although we cannot change our genes, we can maximize their effectiveness by shaping their environment.[3]

The Health, Risk Factors, Exercise Training, and Genetics (HERITAGE) family study measured the VO_{2max} of more than 700 subjects before and after twenty months of supervised exercise training. Exercise sessions started with thirty minutes of exercise at 55 percent of VO_{2max} and built to 50 minutes at 75 percent VO_{2max}. Responses to training differed dramatically across subjects in the HERITAGE study. Some "nonresponders" gained no measurable benefit from months of workouts. Others enjoyed huge performance boosts, expanding their aerobic capacities up to 50 percent above starting levels. As with VO_{2max} before training, about half of the training response was attributable to genetics. But the benefits of training did not correlate with starting fitness levels—some people with low starting VO_{2max} showed big gains; others did not. Thus certain genes probably contribute to the training response but not to baseline VO_{2max}, whereas other genes contribute to baseline VO_{2max} but not to training. In summary, genetic differences between people affect how they respond to training, and these genetic differences become apparent only after training—a clear example of the influence of environment on genes.[4]

Scientists collected DNA from participants in the HERITAGE study and are now probing for links between the participants' genetics and their training responses. Such work represents a first step toward training recommendations that account for athletes' genetic profiles. If we understood the genetic basis of athletes' responses to a variety of different kinds of training programs, we could predict individual reactions and make appropriate recommendations. The genetic profiles of some people—such as the high responders in the HERITAGE study—predispose them to improve with classic aerobic training. People with different genetic profiles may respond better to programs that incorporate interval training or resistance exercise. Coaches, trainers, and health care professionals might use such information to design personalized training programs. Similar approaches could be taken for other environmental factors, such as diet, sleep, supplements, and altitude exposure. By finding the most advantageous match between genes and environment, athletes could make the most of their genetic endowment. Such approaches offer great promise, but challenges remain to be surmounted before they are widely useful for athletes.

Genetics influences exercise even though humans share remarkable genetic similarity with each other. Align the DNA of two people, and their nucleotides match at more than 99 percent of the positions. Moreover, most variability among humans occurs at specific locations on the human genome, places where many humans have one allele—a DNA sequence variant—whereas many other humans have a different allele (see Chapter 3 for more about alleles). These variations, called polymorphisms (literally "many forms"), have leading roles throughout this chapter, so allow me to introduce an analogy that illustrates their properties.

Consider the family sedans built by a particular car manufacturer. Each car in a product line shares the same engine, chassis, and interior design. Thus the cars are identical in their main features. However, certain minor features—exterior paint color, the type of tires installed, whether the seats are heated—vary between cars shipped to dealers. The slightly dissimilar sedans sitting on a car lot resemble the marginally different genomes of people. All human genomes share the same nucleotide at most of their three billion sites, but they vary at defined polymorphic locations, just as the sedans differ only in a few optional features.

Although polymorphic locations on the human genome resemble car options, the sequence variants—alleles—at polymorphic sites resemble the different choices available to buyers for each optional feature. For an optional feature such as paint color, buyers typically choose among only a few possibilities, such as racing red, sleek silver, and gaudy gold. And though a company may make more red cars than gold ones, it makes many cars of each color. Likewise, at each polymorphic site, two or more alleles exist in human populations. At some polymorphisms, two different alleles exist at approximately the same frequency. In other cases, one allele is much more common than others. By definition, polymorphic alleles are found in at least 1 percent of the population, in contrast with rare alleles expressed by only a very small portion of people. Such rare mutations resemble alterations to cars beyond the normal options list. When such variations exist, they might represent a manufacturing defect or the special request of some auto aficionado. Similarly, rare mutations sometimes cause disease or could account for the occasional superathlete.

Let's extend the auto analogy a bit further: if we want to know why some versions of the family sedan sell better than others, we can ignore features that are invariant among the sedans. Nor would it make sense to focus on very rare differences because these can only explain a small portion of buying decisions. Instead, we should determine which optional features attract buyers—these are analogous to desirable genetic variations that correlate with high exercise capacity. If we want to understand the genetic contributions to exercise performance, we can ignore the invariant portions of the human genome and focus on polymorphisms, the most common variations.

Most polymorphisms are single nucleotide substitutions. For example, in Chapter 3, we explored the functional consequences of swapping C for T at one position of the gene encoding actinin-3 (ACTN3). Both the C and T alleles appear at higher than 1 percent frequency in human populations, so these are polymorphic alleles. Single nucleotide polymorphisms (SNPs, pronounced "snips") occur at approximately one in every 300 nucleotides and can be within or outside the portion of a gene that codes for protein. Polymorphisms also include insertions, inversions, and deletions of portions of the genome.[5]

Scientists expect to find more than ten million polymorphic sites, mostly SNPs, scattered across the three billion bases of the human genome. So even though humans have the same nucleotide at most sites, there still are too many polymorphisms to explore in detail. Most genes contain more than one polymorphism, and most of these polymorphisms are unrelated to exercise. How do exercise scientists decide where to start looking for the genetic basis of exercise? Candidate genes offer one way to narrow the scope of our inquiries.

In the candidate gene approach, scientists select genes to study based on their known physiological roles and involvement in disease processes. For example, ACTN3 came to the attention of exercise scientists after clinicians first examined its potential role in congenital muscle diseases.[6] Exercise scientists examined the gene encoding creatine kinase (CKMM) because of the enzyme's central role in energy metabolism (Chapter 2). Many other well-studied candidate genes code for proteins that we explored in previous chapters, including uncoupling proteins (UCP), hypoxia-inducible factors (HIF), peroxisome proliferator-activated receptors (PPAR), and vascular endothelial growth factors (VEGF). Scientists have probed the association of many of these genes to measures of exercise performance.

To see how scientists test for links between genes and performance, let's look at the gene for angiotensin-converting enzyme (ACE), a well-studied gene that we have not yet discussed.[7] ACE activates angiotensin—a hormone that regulates blood pressure and muscle growth—by converting it to an active form. Two decades ago, scientists discovered a polymorphism in the human ACE gene. One version of the gene—the insertion (I) allele—contains an extra stretch of DNA that is absent in the deletion (D) allele. People carrying the D allele have higher ACE activity than those carrying the I allele.[8]

A 1998 paper reported that the ACE I allele appears more frequently in a group of world-class climbers compared to the general population—the first link between genetic variation and athletic prowess. Since that publication, dozens of papers have examined the contribution of the ACE I/D polymorphism to athletic performance, many linking the I allele to endurance. For example, the fastest finishers of the South African Ironman Triathlon are especially likely to express the I allele. Despite the weight of the evidence indicating that ACE variants do influence performance, a number of studies contradict this conclusion and controversy persists. Studies of other candidate genes have also produced inconsistent findings.[9]

To see why it is so difficult to demonstrate connections between candidate genes and exercise performance, let's return to the car analogy. Asking whether a particular gene affects athletics resembles asking whether a particular option contributes to car buyers' decisions. To address this question, we could survey dealerships to determine whether cars with different versions of the option sell more readily. This approach mirrors the way scientists assess the effects of polymorphisms on athletic performance. To determine whether the hypothetical candidate gene GOLD correlates with performance, scientists would explore whether people with different alleles of GOLD vary in their athletic success.

Our ability to link polymorphisms to athletic prowess depends on their effect size—the magnitude of their influences on athletics. Consider whether exterior color affects car-buying decisions. If one line of cars offers buyers a choice between gunmetal gray and battleship gray, the importance of exterior color may be difficult to discern. On the other hand, if another line of cars offers a choice between racing red and petunia pink, the role of exterior color in car-buying decisions will probably become more apparent. Likewise, polymorphisms that exert minor influences on exercise are more difficult to identify than those that greatly affect performance. Most genes that affect exercise explain less than 3 percent of the total performance variation, making their effects difficult to detect amid all the other influences, including environment factors and the effects of other genes.

Because many genes affect exercise performance, with each gene contributing only a small effect, the influence of other genes could mask GOLD's effect. An Olympic champion might possess an unfavorable allele of GOLD but favorable alleles of other genes, just as a car buyer might purchase a car with an undesirable exterior color if it has a snazzy stereo. Genes also interact with each other. For instance, an allele that causes a hormone to be overproduced may have no effect in a person who also has an allele that causes the hormone's receptor to be inactive. This situation resembles a car with a snazzy stereo and cheapo speakers—the shoddy speakers directly affect how buyers value the stereo.

Environmental factors might also obscure GOLD's influence. A poor athlete might have a favorable allele of GOLD but a lifetime habit of eating chips in front of the TV, in the same way that a buyer might decline to buy a car for reasons completely unrelated to the car itself: an obnoxious salesperson, a dealership's location, or an annoying traffic jam during the test drive. Thus nongenetic (environmental) factors can mask the effects of genes on athletics, just as factors other than a car's features can influence buying decisions. Finally, genes interact with the environment, much as the desirability of certain car options depends on the environment. Heated seats, for example, probably interest Alaskans more than Floridians. Likewise, many polymorphisms only matter under certain environmental conditions. For example, a 2000 study demonstrated that army recruits with two copies of the ACE I allele improve their exercise efficiency to a much greater degree than those with two copies of the D allele, even though the two groups had the same efficiency before training. In other words, the gene only affected exercise efficiency during a particular environmental condition: training.[10]

Small effect sizes, variability, and interactions with other genes and the environment mean that picking out the influence of any one gene is a bit like trying to discern your child's clarinet as the band marches by. Detecting the weak signal of a candidate gene's effect in a noisy background of environmental and genetic variation requires big sample sizes. Convincing studies typically require hundreds, or even thousands, of subjects.

Candidate gene research accelerated in the past decade. About a hundred papers now appear annually reporting on genes related to endurance, up from twenty papers in 2000. Though candidate gene studies have produced many insights, the approach has limitations. To envision these limitations, imagine yourself a newcomer to American football (or cricket, if this is too difficult to conceive) with only a rudimentary understanding of the game. Now suppose that video technology allows you to follow only a single player, perhaps the quarterback or field goal kicker. You might learn quite a bit about football by watching only the quarterback, though much would remain mysterious. Focus on the field goal kicker and your introduction to football will be decidedly incomplete. Try to switch the "playercam" between positions and you are likely to become hopelessly confused.[11]

This predicament resembles that of scientists employing the candidate gene approach. On the positive side, the candidate gene approach delves deeply into a single gene, just as watching a playercam produces detailed information about a particular football position. But single-gene studies cannot fully explain exercise, just as a playercam would not paint a complete picture of football. Moreover, if we must select which genes (or players) to study based on incomplete knowledge, then we will inevitably neglect crucial genes. Thus a more global viewpoint would enrich our understanding.

So, instead, imagine a camera mounted on a blimp, providing an overhead view of the entire football field. From this vantage point, you will miss many details. But you will begin to see how players at different positions interact with each other, and you might identify a position—middle linebacker, perhaps—that warrants evaluation in more detail. Genome-wide surveys, which scientists have only recently applied to exercise, resemble this blimpcam. Such surveys superficially scan the genome, looking for genes associated with exercise performance, enabling scientists to discover novel genes that could be involved in exercise. Genome-wide surveys do not produce definitive answers about individual genes. Instead, they generate candidate genes that scientists then must study more deeply.

Genome-wide surveys attempt to uncover all of the genetic variation associated with athletics. These unbiased approaches try to sample every gene, even those without any obvious connection to exercise. A difficult challenge facing these strategies: most human genetic variation is unrelated to exercise. If correlating a single gene to exercise resembles finding a needle in a haystack, then detecting all the genetic variation related to exercise amid the more than ten million polymorphisms in the human genome resembles locating a bunch of needles cast across a vast prairie.[12]

The brute force approach—sequencing every polymorphic site—encounters severe technical and economic obstacles. But scientists can instead employ a genetic property called linkage to survey the entire genome. A quick explanation of linkage: During the production of gametes (egg and sperm), matching chromosomes align. DNA swaps between these paired chromosomes, scrambling the genetic material. The traded DNA travels in chunks, and thus sites that are physically close on the DNA are more likely to be inherited together compared to regions farther apart. Because of linkage, people who share the same sequence at one site are likely to also share sequence similarity at adjacent sites. One polymorphism can therefore serve as a marker for nearby polymorphisms.

To appreciate how geneticists use linkage, imagine yourself with the following challenge: you know nothing about professional basketball, but want to identify the all-time-best NBA players. Unfortunately, you don't have enough time to survey every player's success and you have access to only one statistic—the total number of NBA titles won by each player. A good strategy would be use part of your time to assess a random sample of players, identify those who won championships, then use your remaining time to evaluate their teammates. For example, you might discover that B. J. Armstrong won three NBA titles with the Chicago Bulls in the early 1990s. Looking at Armstrong's Chicago teammates would reveal two players—Will Perdue and Horace Grant—who won four championships, and two players—Michael Jordan and Scottie Pippen—who won six championships. Without a doubt, you've identified a bunch of players who might be all-time greats. One strategy employed by exercise scientists to uncover genes that contribute to athletic performance resembles this approach. First, scientists evaluate evenly spaced markers—often SNPs—to identify regions associated with performance.

Then, they scour these regions to determine which genes are most likely to be involved.

For example, in a recent report from the HERITAGE study, scientists first correlated the aerobic training response with 701 markers spanning the entire genome. Markers on chromosome 13 correlated highly with responses to training, so the scientists then mapped almost 2,000 SNPs near these markers and identified four genes with strong association to improvements in fitness. One of these genes, coding for a protein called sarcoglycan, had been identified in previous studies, a nice confirmation of the earlier work. In contrast, mitochondrial intermediate peptidase (MIPEP)—the gene most highly correlated with the aerobic training response—had received little attention from exercise physiologists prior to this work, even though MIPEP's function makes it a plausible candidate gene. MIPEP codes for an enzyme that activates mitochondrial enzymes, including those involved in aerobic metabolism, by trimming them into their mature form. Thus this genome-wide study uncovered a novel candidate gene worthy of further study.[13]

Coarse-grained linkage analyses followed by fine-grained follow-up studies look at the entire genome with a wide lens, then focus a microscope on a particular region. Recently, breakthroughs both in sequencing technology and our knowledge of the human genome have made possible fine-grained analysis of sites across the entire genomes. Such studies are labeled association studies to differentiate them from linkage analyses that use many fewer markers. Modern association studies usually evaluate large numbers of SNPs—300,000 to 1,000,000—across the entire genome, often allowing scientists to detect correlations with specific genes in the initial screen.[14]

An association study from the HERITAGE study assessed more than 300,000 SNPs in almost 500 Caucasian subjects, finding thirty-nine genes that correlated with the response to aerobic training. Acyl-CoA synthetase long-chain family member 1 (ACSL1), the gene most highly correlated with training, codes for an enzyme involved in fat metabolism. Because physiological studies show that aerobic training shifts metabolism toward using more fat (Chapter 9), we can easily imagine a role for ACSL1 in training. Other genes highly correlated with aerobic training include two transcription factors, proteins that influence expression of other genes. Again, these genes have a plausible role in training: variation in transcription factors could affect how exercise triggers remodeling of muscle cells. Many of the other identified genes code for proteins with less obvious roles in training, including a brain potassium channel and a neurotransmitter receptor. Thus this study is another example of a genome-wide scan discovering new candidate genes.[15]

Impressively, twenty-one of the thirty-nine SNPs from the association study explain about 50 percent of the variation in the response of aerobic capacity to endurance training. Because environmental factors probably contribute about 50 percent of the variability, these twenty-one polymorphisms apparently account for nearly the entire genetic basis of endurance training. Can we now start confidently making training recommendations based on genes? Is the case closed on which genes contribute to endurance training? Far from it.

Some uncertainty always plagues genome-wide studies. The problem is that the statistical tests used to correlate polymorphisms with athletic prowess carry some chance of error for every polymorphism assessed. Even when scientists use stringent statistical criteria, mistakes inevitably occur when hundreds of thousands of polymorphisms are evaluated. Some of these errors result in false positives—genes that are mistakenly taken to be correlated to exercise prowess. Thus we need independent verification of the involvement of the markers identified in the HERITAGE genome-wide scan. However, when that study's authors performed the same analysis with different subject populations—subjects in the HERITAGE study of African descent and participants in two other training studies—they confirmed only some of the SNPs. Another problem is that few of the twenty-one SNPs match candidates genes previously identified to have roles in aerobic training. For example, none of the SNPs resides on chromosome 13, though we know that genes on that chromosome affect exercise. If the study missed some genes that we know to be involved in training, it probably overlooked others as well. Therefore we need more studies to resolve these discrepancies.

Future studies will benefit from continuing improvements in sequencing technology; scientists may soon be able to sequence a person's entire genome quickly for about $1,000.[16] With these advances in sequencing, the limiting factor may soon become recruiting and making physiological measurements on the thousands of subjects needed to achieve statistical confidence. Consider the resources necessary to perform studies like the HERITAGE family study, in which scientists measured the VO_{2max} of over 700 subjects before and after twenty months of supervised exercise training. Coordinated efforts by multiple research centers will be required.

Simply repeating genome-wide scans, however, will not uncover every gene involved in exercise. Some genes may code for proteins integral to exercise but not vary enough among humans to contribute to differences in performance, just as gas tanks are essential to cars' function but may not influence car buyers' decisions because they are invariant across models. Also, for technical reasons, some genes may resist detection in genome-wide association studies. Thus alternative methods are needed to identify the full spectrum of genes contributing to exercise. Two approaches merit mention at this point.

Variations in expression of genes, rather than in their sequences, may also predict responses to training. Recall that messenger ribonucleic acids (mRNAs) are intermediaries between genes and proteins. In addition to assessing genetic variation, the HERITAGE study also measured about 800 mRNAs before exercise training in a group of young adults, using the chip approach described in Chapter 11. When the study's authors correlated gene expression *before training* with posttraining increase in VO_{2max}, differences in levels of twenty-nine genes strongly predicted the training response. In other words, people who most benefited from training had a unique pattern of gene expression even before training began. These differences may be a result of genetic variation— either in the genes whose levels vary or in regulatory genes such as transcription factors that influence their levels—or environmental causes. Somewhat surprisingly, few of the twenty-nine genes change with training. Thus the baseline levels of these genes, rather than how their expression changes during exercise, affect the training response.

Monitoring mRNAs will probably not be a routine strategy for developing personal exercise prescriptions because collecting mRNA requires a muscle biopsy, an invasive procedure that most people will avoid before training. On the other hand, the strategy effectively uncovers candidate genes worth further study. For example, the study described in the previous paragraph further evaluated the twenty-nine identified genes for sequence variations at SNPs. Sequences of eleven of these twenty-nine genes varied between high and low responders, explaining about half of the genetic variability in the training response. These eleven genes differ in both sequence and abundance between high and low responders. Notably, these genes overlap very little with those identified by the HERITAGE genome-wide association scan, so combining studies of expression levels with gene sequencing apparently unmasks novel genes not revealed by other methods.

Although we must interpret their findings cautiously, animal studies offer another means of linking genes to exercise. Because genetic variation in an animal species may differ from variation among humans, animal studies can expose crucial genes missed by human studies. Additionally, animal studies can employ powerful experimental designs, including approaches that would be unethical in humans. A well-conceived mouse breeding study offers a good example: scientists developed four separate lines of "marathon mice" by selectively mating mice that ran more than their littermates. Intriguingly, although all four marathon mice lines share similar motivation and ability to run, they differ in how they achieve the high-running capacity. Most conspicuously, the hindlimbs of two marathon mice lines sport "mini-muscles" that are substantially smaller than the hindlimbs of controls and other runner lines. I too seem to grow minimuscles—I've been called "chicken legs"—so I sympathize with these mice and take solace in the notion that my tiny muscles might be especially adept at running long distances. Genetic analyses of the mice may not only uncover genes that contribute to the desire and capacity to run, but also point toward the different genetic paths that can lead to these traits.[17]

The approaches we've discussed in this chapter—although extraordinarily powerful strategies for exposing genes that *correlate* with exercise performance—do not prove whether the genes actually *cause* differences in athletic talent between people. Nor do these correlations tell us anything about the role of the genes in training. In the same way, linking Armstrong, Grant, Perdue, Pippen, and Jordan to NBA titles neither proves that they contributed to the Bulls' success nor explains their roles on the court. If we examine only the number of titles they won, we cannot know for sure whether these players were true superstars or rode the bench and watched other players win games. To assess the contributions of each player, we'd need more data—perhaps their points, rebounds, steals, and playing minutes during the playoffs. If available, we'd want to watch tapes of those Bulls' playoff games to discover crucial contributions that don't show up on stat sheets. In other words, we'd try to directly assess the function of each player.[18]

Similarly, as we identify new genes associated with exercise, scientists must probe their functional roles. The necessary experiments address questions that will be familiar from previous chapters: What role does the protein encoded by the gene play during exercise? How does its structure contribute to its function? Does its activity level change? Does exercise or training alter its expression level? What happens when we experimentally alter the amount of the protein in cells?

The ultimate holy grail of exercise physiologists is a complete model of the connections between molecules and motion, a model that incorporates all the molecules involved in exercise, that reflects the functions of each molecule, and that accounts for the interactions among these molecules. Unfortunately, we don't have the scientific equivalent of playercams or blimpcams, so mysteries remain, both about how single molecules contribute to exercise and about how myriad molecules work together during exercise. Though we are far from the loftiest goal, the continued efforts to connect molecules to athletics will not only advance exercise science but also offer practical benefits for athletes, coaches, and health care professionals.

The Next Race

Shortly after crossing the finish line, many exhausted marathoners pledge to never run another 26.2-mile race. A day or two later, these same runners often start scanning the listings for their next race. A similar dynamic occurs among other athletes too. The World Series winners celebrate for a few days, then start planning for next year. The cellar-dwellers begin plotting for next season before the playoffs are over. When athletes fail to meet their goals, the next season offers a chance for redemption. When they succeed, new ambitions beckon.

Scientists work much the same way, responding to both successes and failures with renewed effort. Failures greatly outnumber successes in science, so scientists must learn to adjust when their approaches fail. And successes typically lead not to tidy solutions, but to countless new questions. Science pushes forward, so stay alert to new advances in exercise science that build on the material in this book.

But before moving on, wise athletes and scientists reflect upon what they have learned. In that spirit, let's return to the park that we visited at the beginning of this book and again watch amateur athletes at play. We can now appreciate not only the athletes' graceful actions but also the glorious molecular machinery that drives their movements. While athletes jump and swing, proteins contort and morph. While teammates cooperate, molecules bind to each other and work together. And while players make split-second decisions, proteins and ribonucleic acids (RNAs) regulate and coordinate cellular processes.

As we gaze across the playing fields of the park, we can now envision the molecular events that drive athletic movements. On the soccer field, myosin molecules in a player's calf grab actin filaments and pull, powering a penalty kick. Collagen fibers stretch as a basketball player goes for a layup, storing energy that will help power a leap toward the basket. On a nearby tennis court, sodium channels flip open and snap closed, sending electrical impulses that activate the forearm muscles of a player preparing to hit a volley. Out on the softball fields, neurotransmitters flow across synapses and bind to receptors in a shortstop's cerebral cortex, integrating sensory information about the flight path of a line drive. In the liver of a cyclist, enzymes cradle reactants and contort, encouraging the mobilization of fat and carbohydrate. And in response to the hormone vasopressin, water channels insert into cell membranes of an inline skater's kidney, setting the stage for enhanced retention of fluid and thereby postponing dehydration.

Armed with an understanding of the molecules that power exercise, we can begin to perceive how synchronized molecular events produce coordinated athletic motions. Moreover, progress in exercise physiology increasingly enables us to link improvements in athletic performance to specific molecular processes. On the other hand, exercise science does not point to simple, one-size-fits-all recommendations for healthy exercise or powerful training. The preceding pages are filled with examples of complexity, such as the rich tapestry of mechanisms that regulates our metabolism during exercise. Also, we have repeatedly observed variability among individuals, such as the different responses of people to the same exercise program depending on their age, sex, diet, genetic makeup, prior athletic experiences, and many other factors. For these reasons, we need to consider the latest findings of exercise science in light of our own bodies, experiences, and goals.

I hope you find connections between your athletic pursuits and the molecular world discussed in this book. Best wishes for both joy and success in your next race, contest, or adventure.

Visualizing Protein Structures

Two-dimensional images of molecules fail to convey the intricate structure of proteins and other biomolecules. To fully appreciate the proteins discussed in this book, I encourage you to use the free tools developed by the structural biology community. The Research Collaboratory for Structural Bioinformatics (RCSB) Protein Data Bank (PDB) website compiles structural information for over 80,000 molecules.[1] Protein structures are stored in files that contain coordinates for each amino acid.[2] Several free JAVA applets, including the RCSB Protein Workshop and Jmol, allow users to view and manipulate PDB structures.[3] An app is available for mobile devices, too, though it has more limited functions.[4]

To view a protein structure, type its unique code into the search box on the RCSB PDB homepage. Use the code "1CAG" to view a structure of collagen, a protein we discuss in Chapter 1; use the code "1VRP" to view a structure of creatine kinase, which we discuss in Chapter 2.[5] Or you can search for structures using a protein's name. The page for each molecule displays information about the protein, the abstract of the paper that described its structure, and an image of the protein. Click on one of the links below the image to view the protein in the RCSB Protein Workshop or Jmol. Excellent tutorials and help files exist on the RCSB PDB website, so I bypass the technical details and focus on how readers of this book can use these programs to help understand the molecules we discuss.[6] Each program has its advantages—try them both and see which one you like better.

Both Protein Workshop and Jmol allow users to rotate molecules by clicking and dragging the cursor. Try this maneuver with collagen and creatine kinase. Viewing these molecules from different angles yields different insights into their structures. Many protein structures include small molecules bound to the proteins. For example, in the creatine kinase structure, you can see reactants at the active sites. Rotate the creatine kinase molecules to explore the relationship between the protein and these small molecules.

Both programs offer several different styles for viewing molecules. Explore both proteins using these different views:

1. Ribbon diagrams (called cartoons in Jmol) are the most common way to view proteins. Collagen has a simple structure, so the protein chains appears as threads, but creatine kinase has more complex structure, so beta sheets (called "strands" in Protein Workshop) and alpha helices are visible (Chapter 2). This is an excellent way to get an overall impression of a molecule.

2. Ball-and-stick diagrams present only the atoms and bonds of a protein. These contain less information than ribbon

diagrams but are useful for understanding the basic form of protein chains.

3. Jmol allows users to view a CPK model (named after the chemists Robert Corey, Linus Pauling, and Walter Koltun), which shows the space filled by each atom of the protein, providing a more realistic representation than ball and stick models.

4. Surface diagrams depict a view of the contours of a protein's actual surface. These are the most realistic depictions of the protein's external surface, but sometimes obscure structural features that are buried in a protein's interior.

Both Protein Workshop and Jmol allow users to change the colors of the displayed structure. This can lead not only to some beautiful depictions, but may also enable users to better understand protein structures:

1. For multi-subunit proteins like collagen and creatine kinase, changing each chain of the protein to a different color can make it easier to envision how the subunits interact. For example, when each chain of the collagen triple helix is colored differently, the triple helix structure becomes more apparent.

2. Alpha helices and beta sheets within a chain can also each be given a different color, thereby emphasizing these structures.

3. Amino acids in a protein chain can be colored based on hydrophobicity—their propensity to associate with water molecules (see Chapter 1). This feature is especially useful

for exploring membrane proteins with hydrophobic transmembrane domains (see Chapter 6).

In the endnotes to each chapter, I refer readers to the structures
of proteins we discuss and often provide some tips for exploring
them. Where appropriate, I also refer readers to the Molecule of
the Month feature of the RCSB PDB website. These webpages,
written by David S. Goodsell of the Scripps Research Institute
and the RCSB PDB, offer highly readable introductions to the
structures of many proteins.[7]

Finally, those who get hooked on protein structures should explore Foldit—a website that turns predicting protein structures into
a game. Predicting protein structures from their primary amino
acid sequences remains a daunting challenge in biology, and humans still outperform computers in some circumstances. Foldit
invites users to predict structures of biomedically relevant proteins
and design new proteins with potential uses. Users compete with
each other (as individuals or teams) while helping scientists.[8]

Notes

1. Function Follows Form

1. John G. Landels, *Engineering in the Ancient World* (Berkeley: University of California Press, 1978), 107–111; Werner Soedel and Vernard Foley, "Ancient Catapults," *Scientific American* (March 1979): 150–160.

2. General sources for protein structure are Gregory A. Petsko and Dagmar Ringe, *Protein Structure and Function* (London: New Science Press, 2004); Carl Branden and John Tooze, *Introduction to Protein Structure* (New York: Garland, 1999).

3. Christopher C. Lee and Richard L. Jacobs, "Achilles: The Man, the Myth, the Tendon," *Iowa Orthopaedic Journal* 22 (2002): 108–109.

4. The cross-sectional area (CSA) of the Achilles tendon is ~50 mm^2, whereas the CSA of calf muscle is ~5000–6000 mm^2. Data on Achilles tendon anatomy and function come from R. McNeill Alexander, *The Human Machine* (New York: Columbia University Press, 1992), 74–87; Florian Nickisch, "Anatomy of the Achilles Tendon," in *The Achilles Tendon: Treatment and Rehabilitation,* ed. J. A. Nunley (New York: Springer, 2009), 3–16. Information about calf size can be found in Fulvio Lauretani, Cosimo Roberto Russo, Stefania Bandinelli, Benedetta Bartali, Chiara Cavazzini, Angelo Di Iorio, Anna Maria Corsi, Taina Rantanen, Jack M. Guralnik, and Luigi Ferrucci, "Age-Associated Changes in Skeletal Muscles and Their Effect on Mobility: An Operational Diagnosis of Sarcopenia," *Journal of Applied Physiology* 95 (2003): 1851–1860; Michael S. Conley, Jeanne M. Foley, Lori L. Ploutz-Snyder, Ronald A. Meyer, and Gary A. Dudley, "Effect of Acute Head-Down Tilt on Skeletal Muscle Cross-Sectional Area and Proton Transverse Relaxation Time," *Journal of Applied Physiology* 81 (1996): 1572–1577. Prius specifications are at "Toyota Motor Sales: Toyota Prius 2012—Performance and Specifications," http://www.toyota.com/prius-hybrid/specs.html (accessed March 22, 2012).

5. I use "flex" and "extend" rather than the cumbersome technical terms "dorsiflex" and "plantarflex" in this instance and throughout the book.

6. The protein structure denoted by the Research Collaboratory for Structural Bioinformatics Protein Data Bank's (RCSB's) Protein Data Base

Identification (PDB ID): 1CAG represents a synthetic protein that is very similar to collagen. J. Bella, M. Eaton, B. Brodsky, and H. M. Berman, "Crystal and Molecular Structure of a Collagen-like Peptide at 1.9 Å Resolution," *Science* 266 (1994): 75–81. See instructions in the Appendix for accessing and viewing this structure. To further explore collagen's structure, see David Goodsell, "Molecule of the Month: Collagen," http://dx.doi.org/10.2210/rcsb_pdb/mom_2000_4 (accessed August 8, 2012).

7. General sources for collagen are Jürgen Brinckmann, Holger Notbohm, and P. K. Müller, eds., *Collagen: Primer in Structure, Processing, and Assembly* (Berlin: Springer, 2005); Peter Fratzl, ed., *Collagen: Structure and Mechanics* (New York: Springer, 2008).

8. Gretchen Reynolds nicely surveys the science of stretching in her recent book *The First 20 Minutes: Surprising Science Reveals How We Can Exercise Better, Train Smarter, and Live Longer* (New York: Hudson Street Press, 2012), 25–49. Other sources include David Behm and Anis Chaouachi, "A Review of the Acute Effects of Static and Dynamic Stretching on Performance," *European Journal of Applied Physiology* 111 (2011): 2633–2651; M. P. McHugh and C. H. Cosgrave, "To Stretch or Not to Stretch: The Role of Stretching in Injury Prevention and Performance," *Scandinavian Journal of Medicine & Science in Sports* 20 (2009): 169–181; E. Witvrouw, N. Mahieu, P. Roosen, and P. McNair, "The Role of Stretching in Tendon Injuries," *British Journal of Sports Medicine* 41 (2007): 224–226; and Keitaro Kubo, Hiroaki Kanehisa, Yasuo Kawakami, and Tetsuo Fukunaga, "Influence of Static Stretching on Viscoelastic Properties of Human Tendon Structures in Vivo," *Journal of Applied Physiology* 90 (2001): 520–527.

9. An accessible and entertaining book about biomaterials is Steven Vogel, *Cats' Paws and Catapults: Mechanical Worlds of Nature and People* (New York: Norton, 1998). Data on properties of different materials are from J. Gosline, M. Lillie, E. Carrington, P. Guerette, C. Ortlepp, and K. Savage, "Elastic Proteins: Biological Roles and Mechanical Properties," *Philosophical Transactions of the Royal Society of London. Series B: Biological Sciences* 357 (2002): 121–132.

10. Dennis M. Bramble and Daniel E. Lieberman, "Endurance Running and the Evolution of Homo," *Nature* 432 (2004): 345–352; Alexander, *The Human Machine,* 74–87; and A. A. Biewener, "Tendons and Ligaments: Structure, Mechanical Behavior and Biological Function," in *Collagen: Structure and Mechanics,* ed. P. Fratzl (New York: Springer, 2008), 1–13.

11. Joseph W. Freeman and Frederick H. Silver, "Elastic Energy Storage in Unmineralized and Mineralized Extracellular Matrices (ECMs): A

Comparison between Molecular Modeling and Experimental Measurements," *Journal of Theoretical Biology* 229 (2004): 371–381.

12. Bramble and Lieberman, "Endurance Running," 345–352. See Bernd Heinrich's delightful book *Why We Run: A Natural History* (New York: Ecco, 2002) for an accessible description of the centrality of running for humans.

13. R. McNeill Alexander, *Principles of Animal Locomotion* (Princeton, NJ: Princeton University Press, 2003), 122–125; Christopher McGowan, *A Practical Guide to Vertebrate Mechanics* (Cambridge: Cambridge University Press, 1999), 182–186.

14. Christopher McDougall, *Born to Run: A Hidden Tribe, Superathletes, and the Greatest Race the World Has Never Seen* (New York: Knopf, 2009).

15. Both resting and maximum heart rates vary considerably among individuals, and maximum heart rate declines with age. You can estimate your maximum heart rate using the equation: $HR_{max} = 208 - (0.7 \times age)$. Hirofumi Tanaka, Kevin D. Monahan, and Douglas R. Seals, "Age-Predicted Maximal Heart Rate Revisited," *Journal of the American College of Cardiology* 37 (2001): 153–156.

16. Gosline et al., "Elastic Proteins," 121–132.

17. Bruce Alberts, Alexander Johnson, Julian Lewis, Martin Raff, Keith Roberts, and Peter Walter, *Molecular Biology of the Cell* (New York: Garland Science, 2008), 1189–1191; J. Rosenbloom, W. R. Abrams, and R. Mecham, "Extracellular Matrix 4: The Elastic Fiber," *Federation of American Societies for Experimental Biology Journal* 7 (1993): 1208–1218.

18. Bradley S. Fleenor, Kurt D. Marshall, Jessica R. Durrant, Lisa A. Lesniewski, and Douglas R. Seals, "Arterial Stiffening with Ageing Is Associated with Transforming Growth Factor-b1-Related Changes in Adventitial Collagen: Reversal by Aerobic Exercise," *Journal of Physiology* 588 (2010): 3971–3982; Susan J. Zieman, Vojtech Melenovsky, and David A. Kass, "Mechanisms, Pathophysiology, and Therapy of Arterial Stiffness," *Arteriosclerosis, Thrombosis, and Vascular Biology* 25 (2005): 932–943.

19. Decreased stiffness with aging may seem counterintuitive because range of motion typically decreases with aging. But declines in range of motion are probably caused by factors other than tendon stiffness, such as inflammation and muscle weakness. Sally D. Lark, John G. Buckley, David A Jones, and Anthony J. Sargeant, "Knee and Ankle Range of Motion during Stepping Down in Elderly Compared to Young Men," *European Journal of Applied Physiology* 91 (2004): 287–295.

20. N. D. Reeves, C. N. Maganaris, and M. V. Narici, "Effect of Strength Training on Human Patella Tendon Mechanical Properties of Older Individuals," *Journal of Physiology* 548 (2003): 971–981; Michael Kjaer, "Role of Extracellular Matrix in Adaptation of Tendon and Skeletal Muscle to Mechanical Loading," *Physiological Reviews* 84 (2004): 649–698.

21. Stretching may also benefit older adults, but the mechanisms probably differ from those in younger athletes. In older adults, stretching may improve range of motion, balance, control of movement, and even strength: Thomaz N. Burke, Fabio J. R. Franca, Sarah R. F. de Meneses, Rosa M. R. Pereira, and Amelia P. Marques, "Postural Control in Elderly Women with Osteoporosis: Comparison of Balance, Strengthening and Stretching Exercises; A Randomized Controlled Trial," *Clinical Rehabilitation* 26, no. 11 (2012): 1021–1031; Damian C. Stanziano, Bernard A. Roos, Arlette C. Perry, Shenghan Lai, and Joseph F. Signorile, "The Effects of an Active-Assisted Stretching Program on Functional Performance in Elderly Persons: A Pilot Study," *Journal of Clinical Interventions in Aging* 4 (2009): 115–120.

2. An Experiment of One

1. Note that the energy we're talking about in this instance is technically called "free energy," to indicate that it is energy that can be used by cells and to differentiate it from "heat energy," which cannot be used.

2. Data on ATP stores and use rate are from Jeremy Mark Berg, John L. Tymoczko, and Lubert Stryer, *Biochemistry* (New York: W. H. Freeman, 2007), 435.

3. A more complete equation describing this reaction is: ATP + water → ADP + phosphate + H^+ + energy. However, water and hydrogen ions (H^+) are not important to this discussion, so for simplicity I've omitted them in Equations 1 and 2.

4. To be precise, a phosphoryl group is transferred, but the distinction between phosphate and phosphoryl is not important to our discussion.

5. Rucks occur when the ball is on the ground, usually after a tackle. To learn more about rucks, see the International Rubgy Boards Laws at http://www.irblaws.com/EN/laws/4/16/section/law (accessed December 21, 2012).

6. J. W. Lawson and R. L. Veech, "Effects of pH and Free $Mg^{2}+$ on the K_{eq} of the Creatine Kinase Reaction and Other Phosphate Hydrolyses and Phosphate Transfer Reactions," *Journal of Biological Chemistry* 254 (1979): 6528–6537.

7. Extensive information about enzymes is at "Brenda: The Comprehensive Enzyme Information System," http://www.brenda-enzymes.org (accessed October 21, 2011). The turnover rate for creatine kinase listed on that site is calculated from data in Michael J. McLeish and George L. Kenyon, "Relating Structure to Mechanism in Creatine Kinase," *Critical Reviews in Biochemistry and Molecular Biology* 40 (2005): 1–20.

8. Information about diagnostic tests for creatine kinase can be found at "Medline Plus: Creatine Phosphokinase Test," http://www.nlm.nih.gov /medlineplus/ency/article/003503.htm (accessed June 18, 2012). Sources for creatine kinase testing: Paola Brancaccio, Nicola Maffulli, and Francesco Mario Limongelli, "Creatine Kinase Monitoring in Sport Medicine," *British Medical Bulletin* 81–82 (2007): 209–230; Greg G. Ehlers, Thomas E. Ball, and Linda Liston, "Creatine Kinase Levels Are Elevated during 2-a-Day Practices in Collegiate Football Players," *Journal of Athletic Training* 37 (2002): 151–156; Ulrich Hartmann and Joachim Mester, "Training and Overtraining Markers in Selected Sport Events," *Medicine and Science in Sports and Exercise* 32 (2000): 209–215.

9. Theo Wallimann, Malgorzata Tokarska-Schlattner, and Uwe Schlattner, "The Creatine Kinase System and Pleiotropic Effects of Creatine," *Amino Acids* 40 (2011): 1271–1296.

10. The protein structure denoted by the Research Collaboratory for Structural Bioinformatics Protein Data Bank's (RCSB's) Protein Data Base Identification (PDB ID): 1VRP represents creatine kinase from an electric ray. Sushmita D. Lahiri, Pan-Fen Wang, Patricia C. Babbitt, Michael J. McLeish, George L. Kenyon, and Karen N. Allen, "The 2.1 Å Structure of *Torpedo californica* Creatine Kinase Complexed with the ADP-Mg2+ NO3-Creatine Transition-State Analogue Complex," *Biochemistry* 41 (2002): 13861–13867. See instructions in the Appendix for accessing, viewing, and interpreting this structure. Note that a functional creatine kinase enzyme is formed when two identical creatine kinase subunits assemble together. This structure captures both subunits of the creatine kinase enzyme. Figure 2.2 was drawn with Jmol: an open-source Java viewer for chemical structures in 3D (http://www.jmol.org/).

11. McLeish and Kenyon, "Relating Structure," 1–20.

12. J. A. Mejsnar, B. Sopko, and M. Gregor, "Myofibrillar Creatine Kinase Activity Inferred from a 3D Model," *Physiological Research* 51 (2002): 35–41; Lahiri et al., "The 2.1 Å Structure of *Torpedo californica* Creatine Kinase," 13861–13867.

13. The different conformations of creatine kinase can be seen in the PDB ID: 1VRP structure by rotating the molecules and looking at the relationship between creatine kinase and the small molecules associated with the enzyme. One subunit represents the structure of the enzyme with both reactants bound to the active site—the loops are closed over the active site, forming an enclosed binding pocket that promotes the transition state. The other subunit has only ADP bound at the active site, so the active site remains more open.

14. John S. Cantwell, Walter R. Novak, Pan-Fen Wang, Michael J. McLeish, George L. Kenyon, and Patricia C. Babbitt, "Mutagenesis of Two Acidic Active Site Residues in Human Muscle Creatine Kinase: Implications for the Catalytic Mechanism," *Biochemistry* 40 (2001): 3056–3061.

15. K. Vandenberghe, N. Gillis, M. Van Leemputte, P. Van Hecke, F. Vanstapel, and P. Hespel, "Caffeine Counteracts the Ergogenic Action of Muscle Creatine Loading," *Journal of Applied Physiology* 80 (1996): 452–457.

16. Hundreds of papers have explored the effects of creatine supplementation. The American College of Sports Medicine statement on creatine supplementation is an excellent summary: R. L. Terjung, P. Clarkson, E. R. Eichner, P. L. Greenhaff, P. J. Hespel, R. G. Israel, W. J. Kraemer, R. A. Meyer, L. L. Spriet, M. A. Tarnopolsky, A. J. Wagenmakers, and M. H. Williams, "The Physiological and Health Effects of Oral Creatine Supplementation," *Medicine and Science in Sports and Exercise* 32 (2000): 706–717. Jeffrey R. Stout, Jose Antonio, and Douglas Kalman, *Essentials of Creatine in Sports and Health* (Totowa, NJ: Humana, 2008) summarizes recent work.

17. Caroline Rae, Alison L. Digney, Sally R. McEwan, and Timothy C. Bates, "Oral Creatine Monohydrate Supplementation Improves Brain Performance: A Double-Blind, Placebo-Controlled, Cross-over Trial," *Proceedings of the Royal Society of London. Series B: Biological Sciences* 270 (2003): 2147–2150; Piero Sestili, C. Martinelli, E. Colombo, E. Barbieri, L. Potenza, S. Sartini, and C. Fimognari, "Creatine as an Antioxidant," *Amino Acids* 40 (2011): 1385–1396; Wallimann, Tokarska-Schlattner, and Schlattner, "The Creatine Kinase System," 1271–1296.

18. Bert O. Eijnde, Marc Van Leemputte, Marina Goris, Valery Labarque, Youri Taes, Patricia Verbessem, Luc Vanhees, Monique Ramaekers, Bart Vanden Eynde, Reinout Van Schuylenbergh, Ren Dom, Erik A. Richter, and Peter Hespel, "Effects of Creatine Supplementation and Exercise Training on Fitness in Men 55–75 Yr Old," *Journal of Applied Physiology* 95 (2003): 818–828; E. S. Rawson, P. M. Clarkson, T. B. Price, and M. P. Miles, "Differential Response of Muscle Phosphocreatine to Creatine Sup-

plementation in Young and Old Subjects," *Acta Physiologica Scandinavica* 174 (2002): 57–65; M. A. Tarnopolsky, "Caffeine and Creatine Use in Sport," *Annals of Nutrition and Metabolism* 57 (2010): 1–8; David Benton and Rachel Donohoe, "The Influence of Creatine Supplementation on the Cognitive Functioning of Vegetarians and Omnivores," *British Journal of Nutrition* 105 (2011): 1100–1105.

19. George Sheehan, "Each of Us Is an Expert in the Self," http://georgesheehan.com/expertself.html (accessed October 21, 2011).

3. The Gene for Gold Medals

1. The protein's complete name is alpha-actinin-3. But for simplicity I'll omit the prefix alpha throughout this chapter.

2. For a history of the one gene one enzyme hypothesis, see Michel Morange, "The One Gene–One Enzyme Hypothesis," in *A History of Molecular Biology,* trans. Matthew Cobb (Cambridge, MA: Harvard University Press, 1998), 21–29.

3. Several different protein networks comprise the cytoskeleton. We focus on only one of these systems—the actin cytoskeleton—in this chapter. For information about the rest of the cytoskeleton, see Harvey Lodish, Arnold Berk, Chris A. Kaiser, Monty Krieger, Matthew P. Scott, Anthony Bretscher, Hidde Ploegh, and Paul Matsudaira, *Molecular Cell Biology* (New York: W. H. Freeman, 2008), 713–800.

4. Ibid., 717.

5. The protein structure denoted by the Research Collaboratory for Structural Bioinformatics Protein Data Bank's (RCSB's) Protein Data Base Identification (PDB ID): 1ATN represents an individual actin subunit. Wolfgang Kabsch, Hans Georg Mannherz, Dietrich Suck, Emil F. Pai, and Kenneth C. Holmes, "Atomic Structure of the Actin: DNase I Complex," *Nature* 347 (1990): 37–44. See instructions in the Appendix for accessing and viewing this structure. Note that this structure depicts actin (Chain A) in association with an enzyme (deoxyribonuclease, Chain B). Actin is the larger, globular molecule. To further explore actin's structure, see David Goodsell, "Molecule of the Month: Actin," http://dx.doi.org/10.2210/rcsb_pdb/mom_2001_7 (accessed August 7, 2012).

6. PDB ID: 1SJJ represents the structure of an actinin dimer from chicken smooth muscle. Jun Liu, Dianne W. Taylor, and Kenneth A. Taylor, "A 3-D Reconstruction of Smooth Muscle ∝-actinin by CryoEm Reveals Two Different Conformations at the Actin-Binding Region," *Journal of Molecular Biology* 338 (2004): 115–125. See instructions in the Appendix for

accessing and viewing this structure. Use the surfaces style and color each chain separately to visualize the head-to-tail structure of actinin.

7. For an interesting perspective on unity and diversity in biology, see George N. Somero, "Unity in Diversity: A Perspective on the Methods, Contributions, and Future of Comparative Physiology," *Annual Review of Physiology* 62 (2000): 927–937.

8. Brad R. Moon, J. Johanna Hopp, and Kevin E. Conley, "Mechanical Trade-offs Explain How Performance Increases without Increasing Cost in Rattlesnake Tailshaker Muscle," *Journal of Experimental Biology* 205 (2002): 667–675; Stefan Galler, Julia Litzlbauer, Markus Kröss, and Herbert Grassberger, "The Highly Efficient Holding Function of the Mollusc "Catch" Muscle Is Not Based on Decelerated Myosin Head Cross-Bridge Cycles," *Proceedings of the Royal Society B: Biological Sciences* 277 (2010): 803–808; Stephen M. Deban, James C. O'Reilly, Ursula Dicke, and Johan L. van Leeuwen, "Extremely High-Power Tongue Projection in Plethodontid Salamanders," *Journal of Experimental Biology* 210 (2007): 655–667.

9. The nomenclature of muscle types is complicated and varies across species. Type 2B muscles are described in some species; these have properties similar to type 2X muscles.

10. For an excellent review on sarcomere structure, see Mathias Gautel, "The Sarcomeric Cytoskeleton: Who Picks Up the Strain?," *Current Opinion in Cell Biology* 23 (2011): 39–46; Vincent J. Caiozzo, "The Muscular System: Structural and Functional Plasticity," in *American College of Sports Medicine's Advanced Exercise Physiology,* ed. P. A. Farrell, M. J. Joyner, and V. J. Caiozzo (Baltimore, MD: Lippincott, Williams, and Wilkins, 2012), 124–126.

11. Lodish et al., *Molecular Cell Biology,* 247.

12. Greg Elgar and Tanya Vavouri, "Tuning In to the Signals: Noncoding Sequence Conservation in Vertebrate Genomes," *Trends in Genetics* 24 (2008): 344–352.

13. Yemima Berman and Kathryn N. North, "A Gene for Speed: The Emerging Role of ∝-actinin-3 in Muscle Metabolism," *Physiology* 25 (2010): 250–259.

14. Monkol Lek, Kate G. R. Quinlan, and Kathryn N. North, "The Evolution of Skeletal Muscle Performance: Gene Duplication and Divergence of Human Sarcomeric α-actinins," *BioEssays* 32 (2010): 17–25.

15. Michelle Mills, Nan Yang, Ron Weinberger, Douglas L. Vander Woude, Alan H. Beggs, Simon Easteal, and Kathryn North, "Differential Expression of the Actin-Binding Proteins, ∝-actinin-2 and -3, in Different

Species: Implications for the Evolution of Functional Redundancy," *Human Molecular Genetics* 10 (2001): 1335–1346.

16. Nan Yang, Daniel G. MacArthur, Jason P. Gulbin, Allan G. Hahn, Alan H. Beggs, Simon Easteal, and Kathryn North, "ACTN3 Genotype Is Associated with Human Elite Athletic Performance," *American Journal of Human Genetics* 73 (2003): 627–631.

17. Daniel G. MacArthur, Jane T. Seto, Joanna M. Raftery, Kate G. Quinlan, Gavin A. Huttley, Jeff W. Hook, Frances A. Lemckert, Anthony J. Kee, Michael R. Edwards, Yemima Berman, Edna C. Hardeman, Peter W. Gunning, Simon Easteal, Yang Nan, and Kathryn N. North, "Loss of ACTN3 Gene Function Alters Mouse Muscle Metabolism and Shows Evidence of Positive Selection in Humans," *Nature Genetics* 39 (2007): 1261–1265.

18. Monkol Lek and Kathryn N. North, "Are Biological Sensors Modulated by Their Structural Scaffolds? The Role of the Structural Muscle Proteins α-actinin-2 and α-actinin-3 as Modulators of Biological Sensors," *Federation of European Biochemical Societies Letters* 584 (2010): 2974–2980.

19. Berman and North, "A Gene for Speed," 250–259; MacArthur et al., "Loss of ACTN3 Gene Function," 1261–1265; and Y. Ogura, H. Naito, R. Kakigi, T. Akema, T. Sugiura, S. Katamoto, and J. Aoki, "Different Adaptations of Alpha-Actinin Isoforms to Exercise Training in Rat Skeletal Muscles," *Acta Physiologica* 196 (2009): 341–349.

20. For example, Claude Bouchard, Tuomo Rankinen, Yvon C. Chagnon, Treva Rice, Louis Pérusse, Jacques Gagnon, Ingrid Borecki, Ping An, Arthur S. Leon, James S. Skinner, Jack H. Wilmore, Michael Province, and D. C. Rao, "Genomic Scan for Maximal Oxygen Uptake and Its Response to Training in the HERITAGE Family Study," *Journal of Applied Physiology* 88 (2000): 551–559.

21. Colin N. Moran, Nan Yang, Mark E. S. Bailey, Athanasios Tsiokanos, Athanasios Jamurtas, Daniel G. MacArthur, Kathryn North, Yannis P. Pitsiladis, and Richard H. Wilson, "Association Analysis of the ACTN3 R577X Polymorphism and Complex Quantitative Body Composition and Performance Phenotypes in Adolescent Greeks," *European Journal of Human Genetics* 15 (2006): 88–93; Anastasiya Druzhevskaya, Ildus Ahmetov, Irina Astratenkova, and Viktor Rogozkin, "Association of the ACTN3 R577X Polymorphism with Power Athlete Status in Russians," *European Journal of Applied Physiology* 103 (2008): 631–634.

4. Not Too Fast, Not Too Slow

1. John Jerome, *The Sweet Spot in Time* (New York: Breakaway Books, 1998), 26.

2. Technically, velocity also includes a specification of the motion's direction.

3. Milton Hildebrand, "Digging of Quadrupeds," in *Functional Vertebrate Morphology,* ed. M. Hildebrand, D. M. Bramble, K. F. Liem, and D. B. Wake (Cambridge, MA: Belknap Press of Harvard University Press, 1985), 94–98; Andrew N. Iwaniuk, Sergio M. Pellis, and Ian Q. Whishaw, "The Relationship between Forelimb Morphology and Behaviour in North American Carnivores (Carnivora)," *Canadian Journal of Zoology* 77 (1999): 1064–1074.

4. Karen L. Steudel-Numbers, Timothy D. Weaver, and Cara M. Wall-Scheffler, "The Evolution of Human Running: Effects of Changes in Lower-Limb Length on Locomotor Economy," *Journal of Human Evolution* 53 (2007): 191–196.

5. The protein structure denoted by the Research Collaboratory for Structural Bioinformatics Protein Data Bank's (RCSB's) Protein Data Base Identification (PDB ID): 1B7T represents a partial structure of myosin. Anne Houdusse, Vassilios N. Kalabokis, Daniel Himmel, Andrew G. Szent-Györgyi, and Carolyn Cohen, "Atomic Structure of Scallop Myosin Subfragment S1 Complexed with MgADP: A Novel Conformation of the Myosin Head," *Cell* 97 (1999): 459–470. See instructions in the Appendix for accessing and viewing this structure. Note that this structure depicts only the neck and head of a myosin molecule. To further explore myosin's structure, see David Goodsell, "Molecule of the Month: Myosin," http://dx.doi.org/10.2210/rcsb_pdb/mom_2001_6 (accessed August 7, 2012).

6. H. Lee Sweeney and Anne Houdusse, "Structural and Functional Insights into the Myosin Motor Mechanism," *Annual Review of Biophysics* 39 (2010): 539–557.

7. Gabriella Piazzesi, Massimo Reconditi, Marco Linari, Leonardo Lucii, Pasquale Bianco, Elisabetta Brunello, Valérie Decostre, Alex Stewart, David B. Gore, Thomas C. Irving, Malcolm Irving, and Vincenzo Lombardi, "Skeletal Muscle Performance Determined by Modulation of Number of Myosin Motors rather than Motor Force or Stroke Size," *Cell* 131 (2007): 784–795.

8. The average weight of an amino acid is 110 daltons, so a protein of 100 amino acids weighs 11,000 daltons, or 11 kDa. Most proteins fall between 10 and 200 kDa. The smallest proteins, synthesized in the lab, are

about 2 kDa. Jonathan W. Neidigh, R. Matthew Fesinmeyer, and Niels H. Andersen, "Designing a 20-Residue Protein," *Nature Structure and Molecular Biology* 9 (2002): 425–430.

9. Calculation of number of actin subunits is based on values from Harvey Lodish, Arnold Berk, Chris A. Kaiser, Monty Krieger, Matthew P. Scott, Anthony Bretscher, Hidde Ploegh, and Paul Matsudaira, *Molecular Cell Biology* (New York: W. H. Freeman, 2008), 717. Some recent work contradicts the hypothesis presented in the paragraph. Instead, nebulin may stabilize thin filaments, preventing them from dissociating into individual actin subunits. These two roles are not mutually exclusive; nebulin could both stabilize actin filaments and determine their length. See Christopher T. Pappas, Katherine T. Bliss, Anke Zieseniss, and Carol C. Gregorio, "The Nebulin Family: An Actin Support Group," *Trends in Cell Biology* 21 (2011): 29–37.

10. D. E. Rassier, B. R. MacIntosh, and W. Herzog, "Length Dependence of Active Force Production in Skeletal Muscle," *Journal of Applied Physiology* 86 (1999): 1445–1457.

11. Martina Krüger and Wolfgang A. Linke, "The Giant Protein Titin: A Regulatory Node That Integrates Myocyte Signaling Pathways," *Journal of Biological Chemistry* 286 (2011): 9905–9912.

12. Edward G. Moczydlowski and Michael Apkon, "Cellular Physiology of Skeletal, Cardiac, and Smooth Muscle," in *Medical Physiology,* ed. W. F. Boron and E. L. Boulpaep (Philadelphia: Saunders, 2009), 237–263.

13. R. G. Eston, J. Mickleborough, and V. Baltzopoulos, "Eccentric Activation and Muscle Damage: Biomechanical and Physiological Considerations during Downhill Running," *British Journal of Sports Medicine* 29 (1995): 89–94.

14. Edith M. Arnold and Scott L. Delp, "Fibre Operating Lengths of Human Lower Limb Muscles during Walking," *Philosophical Transactions of the Royal Society B: Biological Sciences* 366 (2011): 1530–1539.

15. M. Ishikawa, J. Pakaslahti, and P. V. Komi, "Medial Gastrocnemius Muscle Behavior during Human Running and Walking," *Gait & Posture* 25 (2007): 380–384; Rassier, MacIntosh, and Herzog, "Length Dependence," 1445–1457.

16. N. McCartney, G. Obminski, and G. J. Heigenhauser, "Torque-Velocity Relationship in Isokinetic Cycling Exercise," *Journal of Applied Physiology* 58 (1985): 1459–1462.

17. Matthew J. Tyska and David M. Warshaw, "The Myosin Power Stroke," *Cell Motility and the Cytoskeleton* 51 (2002): 1–15; Caspar Ruegg, Claudia Veigel, Justin E. Molloy, Stephan Schmitz, John C. Sparrow, and

Rainer H. A. Fink, "Molecular Motors: Force and Movement Generated by Single Myosin II Molecules," *Physiology* 17 (2002): 213–218.

18. Piazzesi et al., "Skeletal Muscle Performance," 784–795; Joe Howard, "Molecular Motors: Structural Adaptations to Cellular Functions," *Nature* 389 (1997): 561–567.

19. Piazzesi et al., "Skeletal Muscle Performance," 784–795; Zhen-He He, Roberto Bottinelli, Maria A. Pellegrino, Michael A. Ferenczi, and Carlo Reggiani, "ATP Consumption and Efficiency of Human Single Muscle Fibers with Different Myosin Isoform Composition," *Biophysical Journal* 79 (2000): 945–961.

20. Howard, "Molecular Motors," 561–567.

21. Enrique M. De La Cruz and E. Michael Ostap, "Relating Biochemistry and Function in the Myosin Superfamily," *Current Opinion in Cell Biology* 16 (2004): 61–67; W. H. Guilford, D. E. Dupuis, G. Kennedy, J. Wu, J. B. Patlak, and D. M. Warshaw, "Smooth Muscle and Skeletal Muscle Myosins Produce Similar Unitary Forces and Displacements in the Laser Trap," *Biophysical Journal* 72 (1997): 1006–1021.

22. Vincent J. Caiozzo, "The Muscular System: Structural and Functional Plasticity," in *American College of Sports Medicine's Advanced Exercise Physiology,* ed. P. A. Farrell, M. J. Joyner, and V. J. Caiozzo (Baltimore, MD: Lippincott, Williams, and Wilkins, 2012), 122.

23. A. T. Bahill and M. M. Freitas, "Two Methods for Recommending Bat Weights," *Annals of Biomedical Engineering* 23 (1995): 436–444.

24. Robert Kemp Adair, *The Physics of Baseball,* 3rd ed. (New York: Perennial, 2002), 112–121; Bahill and Freitas, "Two Methods," 436–444.

5. Lactic Acid Acquitted

1. Both quotes are from the rebuttal sections of Point: Counterpoint debates: Robert A. Robergs, "Counterpoint: Muscle Lactate and H+ Production Do Not Have a 1:1 Association in Skeletal Muscle," *Journal of Applied Physiology* 110 (2011): 1491; Michael I. Lindinger and George J. F. Heigenhauser, "Counterpoint: Lactic Acid Is Not the Only Physicochemical Contributor to the Acidosis of Exercise," *Journal of Applied Physiology* 105 (2008): 361.

2. Kalyan C. Vinnakota and Martin J. Kushmerick, "Point: Muscle Lactate and H+ Production Do Have a 1:1 Association in Skeletal Muscle," *Journal of Applied Physiology* 110 (2011): 1487–1488.

3. The precise yield of ATP molecules depends on many factors, including the amount of ATP that cells need to expend in order to extract glu-

cose's ATP and the efficiency of the cell's mechanisms for converting glucose to ATP. In many instances, the yield may be lower than 32 ATPs per glucose. Chapter 8 discusses uncoupling proteins, molecules that contribute to the variation in ATP yield.

4. The abbreviation VO_2 refers to the volume (V) of oxygen (O_2) consumed per unit time. Sometimes, a dot is placed over the V to denote that VO_2 is a rate of oxygen consumption, not simply a volume.

5. George A. Brooks, "Intra- and Extra-cellular Lactate Shuttles," *Medicine and Science in Sports and Exercise* 32 (2000): 790–799; L. B. Gladden, "Lactate Metabolism: A New Paradigm for the Third Millennium," *Journal of Physiology* 558 (2004): 5–30; Andrew Philp, Adam L. Macdonald, and Peter W. Watt, "Lactate—a Signal Coordinating Cell and Systemic Function," *Journal of Experimental Biology* 208 (2005): 4561–4575.

6. George. A. Brooks, G. E. Butterfield, R. R. Wolfe, B. M. Groves, R. S. Mazzeo, J. R. Sutton, E. E. Wolfel, and J. T. Reeves, "Decreased Reliance on Lactate during Exercise after Acclimatization to 4,300 m," *Journal of Applied Physiology* 71 (1991): 333–341; Gerrit van Hall, "Lactate Kinetics in Human Tissues at Rest and during Exercise," *Acta Physiologica* 199 (2010): 499–508.

7. Ralph J. DeBerardinis, "Is Cancer a Disease of Abnormal Cellular Metabolism? New Angles on an Old Idea," *Genetics in Medicine* 10 (2008): 767–777; Saroj P. Mathupala, Young H. Ko, and Peter L. Pedersen, "Hexokinase-2 Bound to Mitochondria: Cancer's Stygian Link to the 'Warburg Effect' and a Pivotal Target for Effective Therapy," *Seminars in Cancer Biology* 19 (2009): 17–24.

8. Jeremy Mark Berg, John L. Tymoczko, and Lubert Stryer, *Biochemistry* (New York: W. H. Freeman, 2007), 805; P. W. Hochachka and G. B. McClelland, "Cellular Metabolic Homeostasis during Large-Scale Change in ATP Turnover Rates in Muscles," *Journal of Experimental Biology* 200 (1997): 381–386.

9. Irreversible reactions (ones that cannot be driven backward because their products contain much less energy than the reactants) are also often the target of regulation.

10. The protein structure denoted by the Research Collaboratory for Structural Bioinformatics Protein Data Bank's (RCSB's) Protein Data Base Identification (PDB ID): 4PFK represents a bacterial phosphofructokinase. P. R. Evans, G. W. Farrants, P. J. Hudson, and H. G. Britton, "Phosphofructokinase: Structure and Control," *Philosophical Transactions of the Royal Society of London. Series B, Biological Sciences* 293 (1981): 53–62.

See instructions in the Appendix for accessing and viewing this structure. Note the globular structure of this enzyme and its ADP/ATP-binding sites. To further explore the structure of phosphofructokinase and other glycolytic enzymes, see David Goodsell, "Molecule of the Month: Glycolytic Enzymes," http://dx.doi.org/10.2210/rcsb_pdb/mom_2004_2 (accessed August 7, 2012).

11. A complication is that ATP is also a reactant of the reaction catalyzes by phosphofructokinase. ATP attaches to two separate sites on phosphofructokinase, reflecting ATP's dual role as reactant and regulator of the enzyme. As a reactant, ATP binds to the enzyme's active site, donates a phosphate, and is converted to ADP. But phosphofructokinase inactivates when ATP sticks to a regulatory site. Thus phosphofructokinase requires some ATP as a reactant, but too much ATP inhibits the enzyme's activity.

12. Adenosine monophosphate (AMP) levels actually have a bigger effect than ADP levels. AMP is formed when ADP levels rise by the following reaction: ADP + ADP ↔ ATP + AMP.

13. Acetate attaches to a molecule called coenzyme A (CoA), which ushers it into the citric acid cycle as acetyl CoA. As its name implies, CoA behaves like an enzyme. Although a much smaller molecule than most enzymes, it works with enzymes to accelerate reactions, but emerges unchanged.

14. Regulation of lactate dehydrogenase—the enzyme that converts pyruvate to lactate—also affects the ratio of fermentation to aerobic metabolism, but for simplicity we focus on pyruvate dehydrogenase.

15. Robert A. Harris, Melissa M. Bowker-Kinley, Boli Huang, and Pengfei Wu, "Regulation of the Activity of the Pyruvate Dehydrogenase Complex," *Advances in Enzyme Regulation* 42 (2002): 249–259.

16. Although I've used phosphofructokinase to illustrate allosteric regulation and pyruvate dehydrogenase to illustrate covalent modification, the story is actually much more complicated; both enzymes are also controlled by the other form of regulation.

17. C. T. Putman, N. L. Jones, L. C. Lands, T. M. Bragg, M. G. Hollidge-Horvat, and G. J. Heigenhauser, "Skeletal Muscle Pyruvate Dehydrogenase Activity during Maximal Exercise in Humans," *American Journal of Physiology—Endocrinology and Metabolism* 269 (1995): E458–E468.

18. Philp, Macdonald, and Watt, "Lactate—a Signal," 4561–4575.

19. Pyruvate dehydrogenase influences the balance between carbohydrate and fat metabolism. When pyruvate dehydrogenase activity is low,

less pyruvate—which is produced from carbohydrates—flows into the citric acid cycle, so fat metabolism is favored.

20. Robert Acton Jacobs, Peter Rasmussen, Christoph Siebenmann, Víctor Díaz, Max Gassmann, Dominik Pesta, Erich Gnaiger, Nikolai Baastrup Nordsborg, Paul Robach, and Carsten Lundby, "Determinants of Time Trial Performance and Maximal Incremental Exercise in Highly Trained Endurance Athletes," *Journal of Applied Physiology* (2011): 1422–1430; Michael J. Joyner and Edward F. Coyle, "Endurance Exercise Performance: The Physiology of Champions," *Journal of Physiology* 586 (2008): 35–44; E. F. Coyle, A. R. Coggan, M. K. Hopper, and T. J. Walters, "Determinants of Endurance in Well-Trained Cyclists," *Journal of Applied Physiology* 64 (1988): 2622–2630.

21. Benjamin F. Miller, Jill A. Fattor, Kevin A. Jacobs, Michael A. Horning, Sang-Hoon Suh, Franco Navazio, and George A. Brooks, "Metabolic and Cardiorespiratory Responses to 'the Lactate Clamp,'" *American Journal of Physiology—Endocrinology and Metabolism* 283 (2002): E889–E898; Benjamin F. Miller, Michael I. Lindinger, Jill A. Fattor, Kevin A. Jacobs, Paul J. LeBlanc, MyLinh Duong, George J. F. Heigenhauser, and George A. Brooks, "Hematological and Acid-Base Changes in Men during Prolonged Exercise with and without Sodium-Lactate Infusion," *Journal of Applied Physiology* 98 (2005): 856–865; Antony D. Karelis, Mariannick Marcil, François Péronnet, and Phillip F. Gardiner, "Effect of Lactate Infusion on M-Wave Characteristics and Force in the Rat Plantaris Muscle during Repeated Stimulation in Situ," *Journal of Applied Physiology* 96 (2004): 2133–2138.

22. Ole B. Nielsen, Frank de Paoli, and Kristian Overgaard, "Protective Effects of Lactic Acid on Force Production in Rat Skeletal Muscle," *Journal of Physiology* 536 (2001): 161–166; Håkan Westerblad, David G. Allen, and Jan Lännergren, "Muscle Fatigue: Lactic Acid or Inorganic Phosphate the Major Cause?," *Physiology* 17 (2002): 17–21.

23. For a nice explanation of the various causes of fatigue, see Alex Hutchinson, *Which Comes First, Cardio or Weights?* (New York: Harper, 2011), 56–62; Jian Cui, Vernon Mascarenhas, Raman Moradkhan, Cheryl Blaha, and Lawrence I. Sinoway, "Effects of Muscle Metabolites on Responses of Muscle Sympathetic Nerve Activity to Mechanoreceptor(s) Stimulation in Healthy Humans," *American Journal of Physiology—Regulatory, Integrative and Comparative Physiology* 294 (2008): R458–R466.

24. Malcolm Gladwell, *Blink: The Power of Thinking without Thinking* (New York: Little, Brown, 2005).

25. Ralph Beneke, Renate M. Leithäuser, and Oliver Ochentel, "Blood Lactate Diagnostics in Exercise Testing and Training," *International Journal of Sports Physiology & Performance* 6 (2011): 8–24; V. L. Billat, P. Sirvent, G. Py, J. P. Koralsztein, and J. Mercier, "The Concept of Maximal Lactate Steady State: A Bridge between Biochemistry, Physiology and Sport Science," *Sports Medicine* 33 (2003): 407–426.

6. Catch an Edge

1. U. Windhorst, "Muscle Proprioceptive Feedback and Spinal Networks," *Brain Research Bulletin* 73 (2007): 155–202.

2. Uwe Proske and Simon C. Gandevia, "The Kinaesthetic Senses," *Journal of Physiology* 587 (2009): 4139–4146.

3. Most cell membranes contain potassium ion channels that are always open. Flow of potassium ions out of the cell through these channels decreases positive charge inside the cell, contributing to inside negative membrane voltage.

4. The protein structure denoted by the Research Collaboratory for Structural Bioinformatics Protein Data Bank's (RCSB's) Protein Data Base Identification (PDB ID): 2ZXE represents a sodium pump from shark rectal glands. Takehiro Shinoda, Haruo Ogawa, Flemming Cornelius, and Chikashi Toyoshima, "Crystal Structure of the Sodium-Potassium Pump at 2.4 Å Resolution," *Nature* 459 (2009): 446–450. See instructions in the Appendix for accessing and viewing this structure. To ascertain which regions are most likely to be membrane domains, color the amino acids based on hydrophobicity. The most hydrophobic regions are more likely to be embedded in the membrane. To further explore the structure of the sodium pump, see David Goodsell, "Molecule of the Month: Sodium-Potassium Pump," http://dx.doi.org/10.2210/rcsb_pdb/mom_2009_10 (accessed August 7, 2012).

5. A speed of 100 mph is equal to 147 feet/sec, and the distance between the blue line and goal line is 60 feet, so the puck gets to the goalie in less than half a second.

6. The alpha subunit of sodium channels, which is the main structural component, is described in the text and shown in Figure 6.5. Sodium channels also contain a beta subunit that spans the membrane only once. PDB ID: 3RW0 represents a structure of a bacterial sodium channel. Jian Payandeh, Todd Scheuer, Ning Zheng, and William A. Catterall, "The Crystal Structure of a Voltage-Gated Sodium Channel," *Nature* 475 (2011): 353–358. See instructions in the Appendix for accessing and viewing this

structure. To visualize the channel's pore, rotate the molecule. Figure 6.5 (below) was drawn with Jmol: an open-source Java viewer for chemical structures in 3D (http://www.jmol.org/).

7. George R. Dubyak, "Ion Homeostasis, Channels, and Transporters: An Update on Cellular Mechanisms," *Advances in Physiology Education* 28 (2004): 143–154; W. A. Catterall, "Molecular Properties of Voltage-Sensitive Sodium Channels," *Annual Review of Biochemistry* 55 (1986): 953–985.

8. William A. Catterall, "Ion Channel Voltage Sensors: Structure, Function, and Pathophysiology," *Neuron* 67 (2010): 915–928; Payandeh et al., "The Crystal Structure," 353–358; Frank Yu and William Catterall, "Overview of the Voltage-Gated Sodium Channel Family," *Genome Biology* 4 (2003): 207.

9. Actual smoke signals can be more complex. For example, columns and puffs can be distinguished, resulting in at least two "letters."

10. Reza Shadmehr, Maurice A. Smith, and John W. Krakauer, "Error Correction, Sensory Prediction, and Adaptation in Motor Control," *Annual Review of Neuroscience* 33 (2010): 89–108.

11. Anna Simon, Fiona Shenton, Irene Hunter, Robert W. Banks, and Guy S. Bewick, "Amiloride-Sensitive Channels Are a Major Contributor to Mechanotransduction in Mammalian Muscle Spindles," *Journal of Physiology* 588 (2010): 171–185.

12. V. Reggie Edgerton and Roland R. Roy, "The Nervous System and Movement," in *American College of Sports Medicine's Advanced Exercise Physiology,* ed. P. A. Farrell, M. J. Joyner, and V. J. Caiozzo (Baltimore, MD: Lippincott, Williams, and Wilkins, 2012), 56–61; Gordon M. Shepherd, *Neurobiology* (Oxford: Oxford University Press, 1988), 280–285; Uwe Proske, "What Is the Role of Muscle Receptors in Proprioception?," *Muscle & Nerve* 31 (2005): 780–787.

13. Sarah Michelle Norris, Eric Bombardier, Ian Curtis Smith, Chris Vigna, and Allan Russell Tupling, "ATP Consumption by Sarcoplasmic Reticulum Ca^{2+} Pumps Accounts for 50% of Resting Metabolic Rate in Mouse Fast and Slow Twitch Skeletal Muscle," *American Journal of Physiology— Cell Physiology* 298 (2010): C521–C529; Lawrence C. Rome and Stan L. Lindstedt, "The Quest for Speed: Muscles Built for High-Frequency Contractions," *Physiology* 13 (1998): 261–268; C. Barclay, R. Woledge, and N. Curtin, "Energy Turnover for Ca^{2+} Cycling in Skeletal muscle," *Journal of Muscle Research and Cell Motility* 28 (2007): 259–274.

14. Michael J. McKenna, Jens Bangsbo, and Jean-Marc Renaud, "Muscle K^+, Na^+, and Cl^- Disturbances and Na^+–K^+ Pump Inactivation: Implications

for Fatigue," *Journal of Applied Physiology* 104 (2008): 288–295; T. Clausen, "Role of Na$^+$, K$^+$-Pumps and Transmembrane Na$^+$, K$^+$ Distribution in Muscle Function," *Acta Physiologica* 192 (2008): 339–349.

7. Your Brain on Exercise

1. Larry W. Swanson, "Mapping the Human Brain: Past, Present, and Future," *Trends in Neurosciences* 18 (1995): 471–474.

2. Ibid.

3. Claude Ghez, "Voluntary Movement," in *Principles of Neural Science,* ed. E. R. Kandel, J. H. Schwartz, and T. M. Jessell (New York: McGraw-Hill, 2000), 756–781; Richard W. Hill, Gordon A. Wyse, and Margaret Anderson, *Animal Physiology* (Sunderland, MA: Sinauer Associates, 2012), 520–521.

4. Each sending cell makes an average of 5–10 synapses onto a motor neuron, so 100,000 synapses represent connections from about 10,000–20,000 neurons. Daniel Kernell, "Motoneurons: Synaptic Control," in *The Motoneurone and Its Muscle Fibers* (Oxford: Oxford University Press, 2006), 135–157; C. J. Heckman and R. M. Enoka, "Physiology of the Motor Neuron and the Motor Unit," in *Clinical Neurophysiology of Motor Neuron Diseases,* ed. A. Eisen (Amsterdam: Elsevier, 2004), 119–147.

5. Constance Hammond, *Cellular and Molecular Neurophysiology* (Amsterdam: Elsevier, 2008), 175.

6. The protein structure denoted by the Research Collaboratory for Structural Bioinformatics Protein Data Bank's (RCSB's) Protein Data Base Identification (PDB ID): 2BG9 represents an acetylcholine receptor from an electric ray. Nigel Unwin, "Refined Structure of the Nicotinic Acetylcholine Receptor at 4 Å Resolution," *Journal of Molecular Biology* 346 (2005): 967–989. See instructions in the Appendix for accessing and viewing this structure. Note the five separate subunits that form the pore. You can observe the pore by rotating the molecule. To further explore the structure of the acetylcholine receptor, see David Goodsell, "Molecule of the Month: Acetylcholine Receptor," http://dx.doi.org/10.2210/rcsb_pdb/mom_2005 _11 (accessed August 7, 2012).

7. These positive ions are mainly sodium ions.

8. Motor neurons release 300,000 acetylcholine molecules, and two acetylcholines are needed to open each receptor.

9. R. K. Powers and M. D. Binder, "Input-Output Functions of Mammalian Motoneurons," *Reviews of Physiology, Biochemistry and Pharmacology* 143 (2001): 137–263; Marc D. Binder, Farrel R. Robinson, and Randall K.

Powers, "Distribution of Effective Synaptic Currents in Cat Triceps Surae Motoneurons. VI. Contralateral Pyramidal Tract," *Journal of Neurophysiology* 80 (1998): 241–248.

10. This effect is especially prevalent as membrane potential rises due to excitatory input. In resting cells, Cl⁻ flows at about equal rates into and out of the cell. Although this flow doesn't itself affect membrane voltage, it does tend to "clamp" membrane voltage at the existing level.

11. David Parker, "Variable Properties in a Single Class of Excitatory Spinal Synapse," *Journal of Neuroscience* 23 (2003): 3154–3163.

12. Sten Grillner, "The Motor Infrastructure: From Ion Channels to Neuronal Networks," *Nature Reviews Neuroscience* 4 (2003): 573–586; A. A. Vandervoort, D. G. Sale, and J. Moroz, "Comparison of Motor Unit Activation during Unilateral and Bilateral Leg Extension," *Journal of Applied Physiology* 56 (1984): 46–51; Ole Kjaerulff and Ole Kiehn, "Crossed Rhythmic Synaptic Input to Motoneurons during Selective Activation of the Contralateral Spinal Locomotor Network," *Journal of Neuroscience* 17 (1997): 9433–9447; N. Khodiguian, A. Cornwell, E. Lares, P. A. DiCaprio, and S. A. Hawkins, "Expression of the Bilateral Deficit during Reflexively Evoked Contractions," *Journal of Applied Physiology* 94 (2003): 171–178.

13. J. D. Howard and R. M. Enoka, "Maximum Bilateral Contractions Are Modified by Neurally Mediated Interlimb Effects," *Journal of Applied Physiology* 70 (1991): 306–316.

14. Noradrenalin and adrenalin are also called norepinephrine and epinephrine, respectively. Modulatory synapses use a variety of neurotransmitters. For example, some varieties of acetylcholine, glutamate, and gamma-aminobutyric acid (GABA) receptors are modulatory, whereas other receptors for these neurotransmitters are ionic. So one neurotransmitter can produce varied effects in receiving cells, depending on the types of receptor the cell expresses.

15. C. J. Heckman, Allison S. Hyngstrom, and Michael D. Johnson, "Active Properties of Motoneurone Dendrites: Diffuse Descending Neuromodulation, Focused Local Inhibition," *Journal of Physiology* 586 (2008): 1225–1231; C. J. Heckman, Robert H. Lee, and Robert M. Brownstone, "Hyperexcitable Dendrites in Motoneurons and Their Neuromodulatory Control during Motor Behavior," *Trends in Neurosciences* 26 (2003): 688–695.

16. The connection between noradrenaline signaling and perception of pace is plausible but speculative; other mechanisms may also contribute.

17. For an accessible introduction to our pleasure circuitry, see David J. Linden, *The Compass of Pleasure: How Our Brains Make Fatty Foods, Orgasm,*

Exercise, Marijuana, Generosity, Vodka, Learning, and Gambling Feel So Good (New York: Viking, 2011).

18. Irving Kupfermann, Eric R. Kandel, and Susan Iversen, "Motivational and Additive States," in *Principles of Neural Science,* ed. E. R. Kandel, J. H. Schwartz, and T. M. Jessell (New York: McGraw-Hill, 2000), 998–1013.

19. Henning Boecker, Till Sprenger, Mary E. Spilker, Gjermund Henriksen, Marcus Koppenhoefer, Klaus J. Wagner, Michael Valet, Achim Berthele, and Thomas R. Tolle, "The Runner's High: Opioidergic Mechanisms in the Human Brain," *Cerebral Cortex* 18 (2008): 2523–2531.

20. There are several different cannabinoid receptors. The CB1 cannabinoid receptors—the ones most involved in exercise—were the target of the knockout study described in this paragraph. Cecilia J. Hillard, "Endocannabinoids and Vascular Function," *Journal of Pharmacology and Experimental Therapeutics* 294 (2000): 27–32; Johannes Fuss and Peter Gass, "Endocannabinoids and Voluntary Activity in Mice: Runner's High and Long-Term Consequences in Emotional Behaviors," *Experimental Neurology* 224 (2010): 103–105; A. Dietrich and W. F. McDaniel, "Endocannabinoids and Exercise," *British Journal of Sports Medicine* 38 (2004): 536–541; Sarah Dubreucq, Muriel Koehl, Djoher N. Abrous, Giovanni Marsicano, and Francis Chaouloff, "CB1 Receptor Deficiency Decreases Wheel-Running Activity: Consequences on Emotional Behaviours and Hippocampal Neurogenesis," *Experimental Neurology* 224 (2010): 106–113.

21. The proteins that link adjacent cells are called cadherins, a name that describes their function—calcium dependent adhesion. Jerry W. Rudy, *The Neurobiology of Learning and Memory* (Sunderland, MA: Sinauer Associates, 2008).

22. Kirk I. Erickson, Michelle W. Voss, Ruchika Shaurya Prakash, Chandramallika Basak, Amanda Szabo, Laura Chaddock, Jennifer S. Kim, Susie Heo, Heloisa Alves, Siobhan M. White, Thomas R. Wojcicki, Emily Mailey, Victoria J. Vieira, Stephen A. Martin, Brandt D. Pence, Jeffrey A. Woods, Edward McAuley, and Arthur F. Kramer, "Exercise Training Increases Size of Hippocampus and Improves Memory," *Proceedings of the National Academy of Sciences* 108 (2011): 3017–3022.

23. N. C. Berchtold, N. Castello, and C. W. Cotman, "Exercise and Time-Dependent Benefits to Learning and Memory," *Neuroscience* 167 (2010): 588–597.

24. John J. Ratey, *Spark: The Revolutionary New Science of Exercise and the Brain* (New York: Little, Brown, 2008) is an entertaining overview of the salutary effects of exercise on the brain.

25. Edwin Li and Kalina Hristova, "Receptor Tyrosine Kinase Transmembrane Domains: Function, Dimer Structure and Dimerization Energetics," *Cell Adhesion & Migration* 4 (2010): 249–254.

8. Live High

1. Max Perutz, "Molecular Biology in Cambridge," in *Cambridge Scientific Minds,* ed. P. Harman and S. Mitton (Cambridge: Cambridge University Press, 2002), 323.

2. Before it enters the citric acid cycle, acetate attaches to coenzyme A (CoA), a small molecule that collaborates with enzymes to catalyze reactions.

3. The water formed this way is called metabolic water. Metabolism can be an important source of water during extreme dehydration. Some desert animals, such as the kangaroo rat, can survive entirely without drinking water by exploiting metabolic water.

4. The quoted value is from William D. McArdle, Frank I. Katch, and Victor L. Katch, *Exercise Physiology* (Baltimore, MD: Williams and Wilkins, 1996), 316. Values from other sources point toward an even greater fold increase. Resting muscle oxygen consumption is ~0.2 milliliter 100 gram^{-1} min^{-1}: Masaki Mizuno, Yuichi Kimura, Takashi Iwakawa, Keiichi Oda, Kenji Ishii, Kiichi Ishiwata, Yoshio Nakamura, and Isao Muraoka, "Regional Differences in Blood Flow and Oxygen Consumption in Resting Muscle and Their Relationship during Recovery from Exhaustive Exercise," *Journal of Applied Physiology* 95 (2003): 2204–2210. Peak muscle oxygen consumption is ~35 milliliter 100 gram^{-1} min^{-1}; G. Rådegran, E. Blomstrand, and B. Saltin, "Peak Muscle Perfusion and Oxygen Uptake in Humans: Importance of Precise Estimates of Muscle Mass," *Journal of Applied Physiology* 87 (1999): 2375–2380.

5. M. Harold Laughlin, Ronald J. Korthuis, Dirk J. Duncker, and Robert J. Bache, "Control of Blood Flow to Cardiac and Skeletal Muscles during Exercise," in *Handbook of Physiology, Section 12: Exercise: Regulation and Integration of Multiple Systems,* ed. L. B. Rowell and J. T. Shepherd (New York: Oxford University Press, 1996), 706.

6. This calculation assumes dry air. In humid air, water vapor displaces some of the oxygen, reducing its partial pressure. Torr is a unit for pressure named after Italian scientist Evangelista Torricelli (1608–1647).

7. Jeffrey B. Graham, "Ecological, Evolutionary, and Physical Factors Influencing Aquatic Animal Respiration," *American Zoologist* 30 (1990): 137–146; Richard W. Hill, Gordon A. Wyse, and M. Anderson, *Animal Physiology* (Sunderland, MA: Sinauer Associates, 2008), 538.

8. Emile L. Boulpaep, "Blood," in *Medical Physiology,* ed. W. F. Boron and E. L. Boulpaep (Philadelphia: Saunders, 2009), 448; Perutz, "Molecular Biology in Cambridge," 323.

9. The protein structure denoted by the Research Collaboratory for Structural Bioinformatics Protein Data Bank's (RCSB's) Protein Data Base Identification (PDB ID): 1HHO represents a human hemoglobin. B. Shaanan, "Structure of Human Oxyhaemoglobin at 2.1 Å Resolution," *Journal of Molecular Biology* 171 (1983): 31–59. See instructions in the Appendix for accessing and viewing this structure. Note the four separate subunits, each with a heme group. To further explore the structure of hemoglobin, see David Goodsell, "Molecule of the Month: Hemoglobin," http://dx.doi.org/10.2210/rcsb_pdb/mom_2003_5 (accessed August 7, 2012).

10. Max F. Perutz, "Molecular Anatomy and Physiology of Hemoglobin," in *Disorders of Hemoglobin,* ed. M. H. Steinberg, B. G. Forget, D. R. Higgs, and R. L. Nagel (Cambridge: Cambridge University Press, 2001), 174–196.

11. Sjaak Philipsen and William G. Wood, "Erythropoiesis," in *Disorders of Hemoglobin: Genetics, Pathophysiology, and Clinical Management,* ed. M. H. Steinberg, B. G. Forget, D. R. Higgs, and D. J. Weatherall (Cambridge: Cambridge University Press, 2009), 24.

12. Benjamin D. Levine, "VO2max: What Do We Know, and What Do We Still Need to Know?," *Journal of Physiology* 586 (2008): 25–34.

13. H. F. Lodish, D. J. Hilton, U. Klingmüller, S. S. Watowich, and H. Wu, "The Erythropoietin Receptor: Biogenesis, Dimerization, and Intracellular Signal Transduction," *Cold Spring Harbor Symposia on Quantitative Biology* 60 (1995): 93–104; Ursula Klingmüller, Ulrike Lorenz, Lewis C. Cantley, Benjamin G. Neel, and Harvey F. Lodish, "Specific Recruitment of SH-PTP1 to the Erythropoietin Receptor Causes Inactivation of JAK2 and Termination of Proliferative Signals," *Cell* 80 (1995): 729–738.

14. Melanie J. Percy, Mary Frances McMullin, Anthony W. W. Roques, Nigel B. Westwood, Jay Acharya, Anne E. Hughes, Terence R. J. Lappin, and Thomas C. Pearson, "Erythrocytosis Due to a Mutation in the Erythropoietin Receptor Gene," *British Journal of Haematology* 100 (1998): 407–410; Albert de la Chapelle, Ann-Liz Traskelin, and Eeva Juvonen, "Truncated Erythropoietin Receptor Causes Dominantly Inherited Benign Human Erythrocytosis," *Proceedings of the National Academy of Sciences of the United States of America* 90 (1993): 4495–4499; Robert Kralovics, Karel Indrak, Tomas Stopka, Brian W. Berman, Jaroslav F. Prchal, and Josef T. Prchal, "Two New EPO Receptor Mutations: Truncated EPO Receptors

Are Most Frequently Associated with Primary Familial and Congenital Polycythemias," *Blood* 90 (1997): 2057–2061.

15. Many RNAs do not code for proteins, but instead have enzymatic or regulatory roles.

16. As we will see in Chapter 11, pre-mRNA is first produced, then processed into mRNA.

17. Sean B. Carroll, *Endless Forms Most Beautiful: The New Science of Evo Devo* (New York: W. W. Norton, 2005), 109–131.

18. Wolfgang Jelkmann, "Regulation of Erythropoietin Production," *Journal of Physiology* 589 (2011): 1251–1258; and Volker H. Haase, "Hypoxic Regulation of Erythropoiesis and Iron Metabolism," *American Journal of Physiology—Renal Physiology* 299 (2010): F1–F13.

19. Jon Peter Wehrlin, Peter Zuest, Jostein Hallén, and Bernard Marti, "Live High-Train Low for 24 Days Increases Hemoglobin Mass and Red Cell Volume in Elite Endurance Athletes," *Journal of Applied Physiology* 100 (2006): 1938–1945; Christoph Siebenmann, Paul Robach, Robert A. Jacobs, Peter Rasmussen, Nikolai Nordsborg, Victor Diaz, Andreas Christ, Niels Vidiendal Olsen, Marco Maggiorini, and Carsten Lundby, " 'Live High–Train Low' Using Normobaric Hypoxia: A Double-Blinded, Placebo-Controlled Study," *Journal of Applied Physiology* 112 (2012): 106–117; Robert F. Chapman, James Stray-Gundersen, and Benjamin D. Levine, "Individual Variation in Response to Altitude Training," *Journal of Applied Physiology* 85 (1998): 1448–1456.

20. Estimates of aerobic metabolism's efficiency vary among sources. The differences in these estimates depend on two factors: the number of adenosine triphosphate (ATP) molecules produced per glucose and the amount of energy required to make ATP from adenosine diphosphate and phosphate. The 50–60 percent estimate was calculated from values in Jeremy Mark Berg, John L. Tymoczko, and Lubert Stryer, *Biochemistry* (New York: W. H. Freeman, 2007), 430 and 455. A higher value (about 70 percent) is reported in Hill, Wyse, and Anderson, *Animal Physiology,* 145.

21. A structure of the ATP synthase is in two PDB files ID: 1C17 (the membrane domain) and ID: 1E79 (the mitochondrial interior domain). Vinit K. Rastogi and Mark E. Girvin, "Structural Changes Linked to Proton Translocation by Subunit c of the ATP Synthase," *Nature* 402 (1999): 263–268; Clyde Gibbons, Martin G. Montgomery, Andrew G. W. Leslie, and John E. Walker, "The Structure of the Central Stalk in Bovine F1-ATPase at 2.4 Å Resolution," *Nature Structural and Molecular Biology* 7 (2000): 1055–1061. See instructions in the Appendix for accessing and viewing

these structures. Observe the differences in hydrophobicity of the membrane and mitochondrial interior domains. To further explore the structure of ATP synthase, see David Goodsell, "Molecule of the Month: ATP synthase," http://dx.doi.org/10.2210/rcsb_pdb/mom_2005_12 (accessed August 7, 2012).

22. Mayumi Nakanishi-Matsui, Mizuki Sekiya, Robert K. Nakamoto, and Masamitsu Futai, "The Mechanism of Rotating Proton Pumping ATPases," *Biochimica et Biophysica Acta—Bioenergetics* 1797 (2010): 1343–1352; Daichi Okuno, Ryota Iino, and Hiroyuki Noji, "Rotation and Structure of F_0F_1-ATP Synthase," *Journal of Biochemistry* 149 (2011): 655–664.

23. Aaron M. Cypess, Sanaz Lehman, Gethin Williams, Ilan Tal, Dean Rodman, Allison B. Goldfine, Frank C. Kuo, Edwin L. Palmer, Yu-Hua Tseng, Alessandro Doria, Gerald M. Kolodny, and C. Ronald Kahn, "Identification and Importance of Brown Adipose Tissue in Adult Humans," *New England Journal of Medicine* 360 (2009): 1509–1517; Barbara Cannon and J. A. N. Nedergaard, "Brown Adipose Tissue: Function and Physiological Significance," *Physiological Reviews* 84 (2004): 277–359; Kirsi A. Virtanen, Martin E. Lidell, Janne Orava, Mikael Heglind, Rickard Westergren, Tarja Niemi, Markku Taittonen, Jukka Laine, Nina-Johanna Savisto, Sven Enerbäck, and Pirjo Nuutila, "Functional Brown Adipose Tissue in Healthy Adults," *New England Journal of Medicine* 360 (2009): 1518–1525.

24. Wouter D. van Marken Lichtenbelt and Patrick Schrauwen, "Implications of Nonshivering Thermogenesis for Energy Balance Regulation in Humans," *American Journal of Physiology—Regulatory, Integrative and Comparative Physiology* 301 (2011): R285–R296.

25. Yong-Jun Liu, Peng-Yuan Liu, Jirong Long, Yan Lu, Leo Elze, Robert R. Recker, and Hong-Wen Deng, "Linkage and Association Analyses of the UCP3 Gene with Obesity Phenotypes in Caucasian Families," *Physiological Genomics* 22 (2005): 197–203; Angela M. Hancock, Vanessa J. Clark, Yudong Qian, and Anna Di Rienzo, "Population Genetic Analysis of the Uncoupling Proteins Supports a Role for UCP3 in Human Cold Resistance," *Molecular Biology and Evolution* 28 (2011): 601–614.

26. Sukhbir Singh Dhamrait, Alun G. Williams, Stephen H. Day, James Robert Skipworth, John R. Payne, Michael World, Steve E. Humphries, and Hugh E. Montgomery, "Variation in the Uncoupling Protein 2 and 3 Genes and Human Performance," *Journal of Applied Physiology* 112 (2012): 1122–1127; Maria Fernström, Michail Tonkonogi, and Kent Sahlin, "Effects of Acute and Chronic Endurance Exercise on Mitochondrial Uncoupling in Human Skeletal Muscle," *Journal of Physiology* 554 (2004): 755–763;

A. Russell, G. Wadley, M. Hesselink, G. Schaart, S. Lo, B. Léger, A. Garnham, E. Kornips, D. Cameron-Smith, J. P. Giacobino, P. Muzzin, R. Snow, and P. Schrauwen, "UCP3 Protein Expression Is Lower in Type I, IIa and IIx Muscle Fiber Types of Endurance-Trained Compared to Untrained Subjects," *Pflügers Archiv European Journal of Physiology* 445 (2003): 563–569.

27. Pontus Böstrom, Jun Wu, Mark P. Jedrychowski, Anisha Korde, Li Ye, James C. Lo, Kyle A. Rasbach, Elisabeth Almer Böstrom, Jang Hyun Choi, Jonathan Z. Long, Shingo Kajimura, Maria Cristina Zingaretti, Birgitte F. Vind, Hua Tu, Saverio Cinti, Kurt Hojlund, Steven P. Gygi, and Bruce M. Spiegelman, "A PGC1-α-Dependent Myokine That Drives Brown-Fat-Like Development of White Fat and Thermogenesis," *Nature* 481 (2012): 463–468.

9. Run Like a Woman

1. I evaluated data from two 2011 races, the Charlotte Thunder Road Marathon and the Belfast City Marathon, because data from these races were available in convenient format on the Internet. In both races, men slowed substantially more than women. Combining the two races, women ran the second half an average of 8.6 percent slower than the first, whereas men ran the second half more than 11 percent slower than the first.

2. Exercise physiologists increasingly use the word "sex" to refer to biological traits, rather than "gender," which is more properly used to describe cultural and social traits related to sex. For more detail about terminology as well as a useful history of research into sex differences in physiology, see Britta N. Torgrimson and Christopher T. Minson, "Sex and Gender: What Is the Difference?," *Journal of Applied Physiology* 99 (2005): 785–787.

3. Tatsuya Hayashi, Jørgen F. P. Wojtaszewski, and Laurie J. Goodyear, "Exercise Regulation of Glucose Transport in Skeletal Muscle," *American Journal of Physiology—Endocrinology and Metabolism* 273 (1997): E1039–E1051.

4. Michale Bouskila, Michael F. Hirshman, Jorgen Jensen, Laurie J. Goodyear, and Kei Sakamoto, "Insulin Promotes Glycogen Synthesis in the Absence of GSK3 Phosphorylation in Skeletal Muscle," *American Journal of Physiology—Endocrinology and Metabolism* 294 (2008): E28–E35.

5. T. E. Graham, Z. Yuan, A. K. Hill, and R. J. Wilson, "The Regulation of Muscle Glycogen: The Granule and Its Proteins," *Acta Physiologica* 199 (2010): 489–498.

6. Lawrence L. Spriet, "The Metabolic Systems: Lipid Metabolism," in *American College of Sports Medicine's Advanced Exercise Physiology,* ed. P. A.

Farrell, M. J. Joyner, and V. J. Caiozzo (Baltimore, MD: Lippincott, Williams, and Wilkins, 2012), 392–407.

7. In both cases, these signals ultimately stimulate kinases that activate glucose transporter movement to the membrane. In the case of ATP, a fall in the ATP to AMP (adenosine monophosphate) ratio activates AMP-activated protein kinase. AMP rises when ATP is split to ADP and phosphate because two ADP molecules can be combined to form ATP and AMP.

8. D. L. Costill, E. Coyle, G. Dalsky, W. Evans, W. Fink, and D. Hoopes, "Effects of Elevated Plasma FFA and Insulin on Muscle Glycogen Usage during Exercise," *Journal of Applied Physiology* 43 (1977): 695–699; Niels Jessen and Laurie J. Goodyear, "Contraction Signaling to Glucose Transport in Skeletal Muscle," *Journal of Applied Physiology* 99 (2005): 330–337; Jack H. Wilmore, David. L. Costill, and W. Larry Kenney, "Body Composition and Nutrition for Sport," in *Physiology of Sport and Exercise* (Champaign, IL: Human Kinetics, 2008), 316–353; Mark Hargreaves, John A. Hawley, and Asker Jeukendrup, "Pre-exercise Carbohydrate and Fat Ingestion: Effects on Metabolism and Performance," *Journal of Sports Sciences* 22 (2004): 31–38; Juul Achten and Asker Jeukendrup, "Effects of Pre-exercise Ingestion of Carbohydrate on Glycaemic and Insulinaemic Responses during Subsequent Exercise at Differing Intensities," *European Journal of Applied Physiology* 88 (2003): 466–471.

9. D. A. Wright, W. M. Sherman, and A. R. Dernbach, "Carbohydrate Feedings before, during, or in Combination Improve Cycling Endurance Performance," *Journal of Applied Physiology* 71 (1991): 1082–1088; J. L. Ivy, A. L. Katz, C. L. Cutler, W. M. Sherman, and E. F. Coyle, "Muscle Glycogen Synthesis after Exercise: Effect of Time of Carbohydrate Ingestion," *Journal of Applied Physiology* 64 (1988): 1480–1485; John L. Ivy, Harold W. Goforth, Bruce M. Damon, Thomas R. McCauley, Edward C. Parsons, and Thomas B. Price, "Early Postexercise Muscle Glycogen Recovery Is Enhanced with a Carbohydrate-Protein Supplement," *Journal of Applied Physiology* 93 (2002): 1337–1344.

10. A. Pottier, J. Bouckaert, W. Gilis, T. Roels, and W. Derave, "Mouth Rinse but Not Ingestion of a Carbohydrate Solution Improves 1-h Cycle Time Trial Performance," *Scandinavian Journal of Medicine & Science in Sports* 20 (2010): 105–111; Asker E. Jeukendrup, "Carbohydrate Intake during Exercise and Performance," *Nutrition* 20 (2004): 669–677; Tom Vandenbogaerde and Will Hopkins, "Effects of Acute Carbohydrate Supplementation on Endurance Performance: A Meta-analysis," *Sports Medicine* 41 (2011): 773–792; Gareth A. Wallis, Ruth Dawson, Juul Achten, Jonathan Webber,

and Asker E. Jeukendrup, "Metabolic Response to Carbohydrate Ingestion during Exercise in Males and Females," *American Journal of Physiology—Endocrinology and Metabolism* 290 (2006): E708–E715. Specific recommendations for athletes can be found in this position statement: American Dietetic Association, Dietitians of Canada, and American College of Sports Medicine, "Nutrition and Athletic Performance," *Medicine & Science in Sports & Exercise* 41 (2009): 709–731.

11. Julien Delezie and Etienne Challet, "Interactions between Metabolism and Circadian Clocks: Reciprocal Disturbances," *Annals of the New York Academy of Sciences* 1243 (2011): 30–46; Ryosuke Doi, Katsutaka Oishi, and Norio Ishida, "CLOCK Regulates Circadian Rhythms of Hepatic Glycogen Synthesis through Transcriptional Activation of Glycogen Synthase 2," *Journal of Biological Chemistry* (2010): 22114–22121.

12. One cup of cooked white rice weighs about 165 g and contains about 50 g carbohydrate (much of the remaining weight is water). Consuming 600 g carbohydrate thus requires eating almost 2,000 g of rice, or about 4.5 lbs.

13. John A. Hawley, Elske J. Schabort, Timothy D. Noakes, and Steven C. Dennis, "Carbohydrate-Loading and Exercise Performance: An Update," *Sports Medicine* 24 (1997): 73–81.

14. M. A. Tarnopolsky, S. A. Atkinson, S. M. Phillips, and J. D. MacDougall, "Carbohydrate Loading and Metabolism during Exercise in Men and Women," *Journal of Applied Physiology* 78 (1995): 1360–1368; Jessica L. Andrews, Darlene A. Sedlock, Michael G. Flynn, James W. Navalta, and Hongguang Ji, "Carbohydrate Loading and Supplementation in Endurance-Trained Women Runners," *Journal of Applied Physiology* 95 (2003): 584–590; J. Lynne Walker, George J. F. Heigenhauser, Eric Hultman, and Lawrence L. Spriet, "Dietary Carbohydrate, Muscle Glycogen Content, and Endurance Performance in Well-Trained Women," *Journal of Applied Physiology* 88 (2000): 2151–2158.

15. In two studies that show expanded glycogen reserves in women, one demonstrates a moderate performance gain; the other did not measure performance; Mark A. Tarnopolsky, Carol Zawada, Lindsay B. Richmond, Sherry Carter, Jane Shearer, Terry Graham, and Stuart M. Phillips, "Gender Differences in Carbohydrate Loading Are Related to Energy Intake," *Journal of Applied Physiology* 91 (2001): 225–230; Anthony P. James, Michael Lorraine, Daelyn Cullen, Carmel Goodman, Brian Dawson, Norman T. Palmer, and Paul A. Fournier, "Muscle Glycogen Supercompensation: Absence of a Gender-Related Difference," *European Journal of Applied Physiology* 85 (2001): 533–538.

16. According to the studies cited in the last two notes, women need to ingest ~12 g carbohydrate per kilogram (kg) fat-free body mass per day to expand glycogen reserves to the same extent that 9 g carbohydrate per kg fat-free body mass per day does in men.

17. K. J. Acheson, Y. Schutz, T. Bessard, K. Anantharaman, J. P. Flatt, and E. Jequier, "Glycogen Storage Capacity and De Novo Lipogenesis during Massive Carbohydrate Overfeeding in Man," *American Journal of Clinical Nutrition* 48 (1988): 240–247; J. P. Flatt, "Use and Storage of Carbohydrate and Fat," *American Journal of Clinical Nutrition* 61 (1995): 952S–959S.

18. As in aerobic metabolism, coenzyme A (CoA), a small molecule that collaborates with enzymes to catalyze reactions, attaches to both acetate and malonate.

19. Fatty-acid synthase has ten distinct regions. One region binds reactants but does not catalyze reactions. Some of the other sites are inactive enzymes. Marc Leibundgut, Timm Maier, Simon Jenni, and Nenad Ban, "The Multienzyme Architecture of Eukaryotic Fatty Acid Synthases," *Current Opinion in Structural Biology* 18 (2008): 714–725.

20. The protein structure denoted by the Research Collaboratory for Structural Bioinformatics Protein Data Bank's (RCSB's) Protein Data Base Identification (PDB ID): 2CF2 represents a pig fatty-acid synthase enzyme. Timm Maier, Simon Jenni, and Nenad Ban, "Architecture of Mammalian Fatty Acid Synthase at 4.5 Å Resolution," *Science* 311 (2006): 1258–1262. See instructions in the Appendix for accessing and viewing this structure. The domains appear as separate chains in this structure, although they are part of the same protein chain in the intact enzymes. To further explore the structure of fatty acid synthase, see David Goodsell, "Molecule of the Month: Fatty Acid Synthase," http://dx.doi.org/10.2210/rcsb_pdb/mom _2007_6 (accessed August 7, 2012).

21. Anne Radenne, Murielle Akpa, Caroline Martel, Sabine Sawadogo, Daniel Mauvoisin, and Catherine Mounier, "Hepatic Regulation of Fatty Acid Synthase by Insulin and T3: Evidence for T3 Genomic and Nongenomic Actions," *American Journal of Physiology—Endocrinology and Metabolism* 295 (2008): E884–E894; Bettina Mittendorfer, "Sexual Dimorphism in Human Lipid Metabolism," *Journal of Nutrition* 135 (2005): 681–686.

22. P. J. Scarpace, M. Matheny, B. H. Pollock, and N. Tumer, "Leptin Increases Uncoupling Protein Expression and Energy Expenditure," *American Journal of Physiology—Endocrinology and Metabolism* 273 (1997): E226–E230; José Donato Jr., Roberta M. Cravo, Renata Frazão, and Carol

F. Elias, "Hypothalamic Sites of Leptin Action Linking Metabolism and Reproduction," *Neuroendocrinology* 93 (2011): 9–18.

23. Information on eating disorders for athletes, coaches, and parents can be found at National Eating Disorders Association, http://www.natio naleatingdisorders.org/ (accessed February 21, 2012). Loretta DiPietro and Nina Stachenfeld have eloquently argued that concerns about associations between eating disorders and athletics should not result in policies that "counter the US Public Health Service efforts to promote the benefits of athletic participation and an active lifestyle among children and adolescents." L. DiPietro, N. S. Stachenfeld, "The Myth of the Female Athlete Triad," *British Journal of Sports Medicine* 40 (2006): 490–493.

24. "Badwater Ultramarathon: Badwater Ultramarathon Race Results, 2000–2011," http://www.badwater.com/results/index.html (accessed February 21, 2012); David P. Speechly, Sheila R. Taylor, and Geoffrey G. Rogers, "Differences in Ultra-endurance Exercise in Performance-Matched Male and Female Runners," *Medicine & Science in Sports & Exercise* 28 (1996): 359–365.

25. Luc J. C. van Loon, "Use of Intramuscular Triacylglycerol as a Substrate Source during Exercise in Humans," *Journal of Applied Physiology* 97 (2004): 1170–1187; Mark A. Tarnopolsky, "Sex Differences in Exercise Metabolism and the Role of 17-beta Estradiol," *Medicine & Science in Sports & Exercise* 40 (2008): 648–654; Carsten Roepstorff, Morten Donsmark, Maja Thiele, Bodil Vistisen, Greg Stewart, Kristian Vissing, Peter Schjerling, D. Grahame Hardie, Henrik Galbo, and Bente Kiens, "Sex Differences in Hormone-Sensitive Lipase Expression, Activity, and Phosphorylation in Skeletal Muscle at Rest and during Exercise," *American Journal of Physiology—Endocrinology and Metabolism* 291 (2006): E1106–E1114.

26. Ted W. Zderic, Andrew R. Coggan, and Brent C. Ruby, "Glucose Kinetics and Substrate Oxidation during Exercise in the Follicular and Luteal Phases," *Journal of Applied Physiology* 90 (2001): 447–453; Tracy J. Horton, Emily K. Miller, Deborah Glueck, and Kathleen Tench, "No Effect of Menstrual Cycle Phase on Glucose Kinetics and Fuel Oxidation during Moderate-Intensity Exercise," *American Journal of Physiology—Endocrinology and Metabolism* 282 (2002): E752–E762; Ming-hua H. Fu, Amy C. Maher, Mazen J. Hamadeh, Changhua Ye, and Mark A. Tarnopolsky, "Exercise, Sex, Menstrual Cycle Phase, and 17ß-Estradiol Influence Metabolism-Related Genes in Human Skeletal Muscle," *Physiological Genomics* 40 (2009): 34–47.

27. Chunyan Zhao, Karin Dahlman-Wright, and Jan-Åke Gustafsson, "Estrogen Signaling via Estrogen Receptor ß," *Journal of Biological Chemistry* 285 (2010): 39575–39579; Amy C. Maher, Mahmood Akhtar, and Mark A. Tarnopolsky, "Men Supplemented with 17ß-Estradiol Have Increased ß-Oxidation Capacity in Skeletal Muscle," *Physiological Genomics* 42 (2010): 342–347; Linda J. Woodhouse, Nidhi Gupta, Meenakshi Bhasin, Atam B. Singh, Robert Ross, Jeffrey Phillips, and Shalender Bhasin, "Dose-Dependent Effects of Testosterone on Regional Adipose Tissue Distribution in Healthy Young Men," *Journal of Clinical Endocrinology & Metabolism* 89 (2004): 718–726.

10. Drinking Games

1. Samuel Mettler, Carmen Rusch, and Paolo C. Colombani, "Osmolality and pH of Sport and Other Drinks Available in Switzerland," *Schweizerische Zeitschrift für Sportmedizin und Sporttraumatologie* 54 (2006): 92–95.

2. G. E. Vist and R. J. Maughan, "The Effect of Osmolality and Carbohydrate Content on the Rate of Gastric Emptying of Liquids in Man," *Journal of Physiology* 486 (1995): 523–531; R. J. Maughan and J. B. Leiper, "Limitations to Fluid Replacement during Exercise," *Canadian Journal of Applied Physiology* 24 (1999): 173–187; T. J. Little, A. Gopinath, E. Patel, A. McGlone, D. J. Lassman, M. D'amato, J. T. McLaughlin, and D. G. Thompson, "Gastric Emptying of Hexose Sugars: Role of Osmolality, Molecular Structure and the CCK1 Receptor," *Neurogastroenterology & Motility* 22 (2010): 1183–1190; Robert Murray, William Bartoli, John Stofan, Mary Horn, and Dennis Eddy, "A Comparison of the Gastric Emptying Characteristics of Selected Sports Drinks," *International Journal of Sport Nutrition* 9 (1999): 263–274.

3. Henry J. Binder, "Organization of the Gastrointestinal Gystem," in *Medical Physiology,* ed. W. F. Boron and E. L. Boulpaep (Philadelphia: Saunders, 2009); World Health Organization, "Cholera Fact Sheet," http://www.who.int/mediacentre/factsheets/fs107/en/index.html (accessed May 2, 2012).

4. Some recent work suggests that water may be carried directly by the sodium-glucose transporter, rather than following solute across epithelial membranes in a separate step. A. K. Meinild, D. A. Klaerke, D. D. F. Loo, E. M. Wright, and T. Zeuthen, "The Human Na$^+$-Glucose Cotransporter Is a Molecular Water Pump," *Journal of Physiology* 508 (1998): 15–21; Ernest M. Wright, Donald D. F. Loo, and Bruce A. Hirayama, "Biology of Human Sodium Glucose Transporters," *Physiological Reviews* 91 (2011):

733–794; Ernest M. Wright, Donald D. F. Loo, Bruce A. Hirayama, and Eric Turk, "Surprising Versatility of Na⁺-Glucose Cotransporters: SLC5," *Physiology* 19 (2004): 370–376.

5. John Stitt, "Regulation of Body Temperature," in *Medical Physiology,* ed. W. F. Boron and E. L. Boulpaep (Philadelphia: Saunders, 2009).

6. Nina G. Jablonski, *Skin: A Natural History* (Berkeley: University of California Press, 2006), 39–55; Daniel Lieberman, *Evolution of the Human Head* (Cambridge, MA: Belknap Press of Harvard University Press, 2011), 209–211.

7. Jablonski, *Skin,* p. 55.

8. Lindsay B. Baker, John R. Stofan, Adam A. Hamilton, and Craig A. Horswill, "Comparison of Regional Patch Collection vs. Whole Body Washdown for Measuring Sweat Sodium and Potassium Loss during Exercise," *Journal of Applied Physiology* 107 (2009): 887–895; Jack H. Wilmore, David L. Costill, and W. Larry Kenney, *Physiology of Sport and Exercise* (Champaign, IL: Human Kinetics, 2008), 262–263.

9. Michael J. Buono, Ryan Claros, Teshina DeBoer, and Janine Wong, "Na⁺ Secretion Rate Increases Proportionally More Than the Na⁺ Reabsorption Rate with Increases in Sweat Rate," *Journal of Applied Physiology* 105 (2008): 1044–1048; Michael J. Buono, Kimberly D. Ball, and Fred W. Kolkhorst, "Sodium Ion Concentration vs. Sweat Rate Relationship in Humans," *Journal of Applied Physiology* 103 (2007): 990–994; C. R. Kirby and V. A. Convertino, "Plasma Aldosterone and Sweat Sodium Concentrations after Exercise and Heat Acclimation," *Journal of Applied Physiology* 61 (1986): 967–970; E. R. Nadel, K. B. Pandolf, M. F. Roberts, and J. A. Stolwijk, "Mechanisms of Thermal Acclimation to Exercise and Heat," *Journal of Applied Physiology* 37 (1974): 515–520.

10. Vasopressin is also called antidiuretic hormone. Cardiovascular physiologists dubbed the hormone vasopressin to reflect its constrictive effect on blood vessels. Renal physiologists named it antidiuretic hormone to reflect its promotion of water retention. Both names have persisted in the literature.

11. Charles W. Bourque and Stéphane H. R. Oliet, "Osmoreceptors in the Central Nervous System," *Annual Review of Physiology* 59 (1997): 601–619; Daniel L. Voisin and Charles W. Bourque, "Integration of Sodium and Osmosensory Signals in Vasopressin Neurons," *Trends in Neurosciences* 25 (2002): 199–205.

12. Gerhard Giebisch and Erich Windhager, "Organization of the Urinary System" and "Glomerular Filtration and Renal Blood Flow," in *Medical*

Physiology, ed. W. F. Boron and E. L. Boulpaep (Philadelphia: Saunders, 2009), chapters 33 and 34.

13. Richard W. Hill, Gordon A. Wyse, and Margaret Anderson, *Animal Physiology* (Sunderland, MA: Sinauer Associates, 2008), 715–748.

14. Water channels are also called aquaporins. The protein structure denoted by the Research Collaboratory for Structural Bioinformatics Protein Data Bank's (RCSB's) Protein Data Base Identification (PDB ID): 3D9S represents a human aquaporin. Rob Horsefield, Kristina Nordén, Maria Fellert, Anna Backmark, Susanna Törnroth-Horsefield, Anke C. Terwisscha van Scheltinga, Jan Kvassman, Per Kjellbom, Urban Johanson, and Richard Neutze, "High-Resolution X-ray Structure of Human Aquaporin 5," *Proceedings of the National Academy of Sciences* 105 (2008): 13327–13332. See instructions in the Appendix for accessing and viewing this structure. Note how four subunits form the pore.

15. Gerhard Giebisch and Erich Windhager, "Integration of Salt and Water Balance," in *Medical Physiology,* ed. W. F. Boron and E. L. Boulpaep (Philadelphia: Saunders, 2009), 866–880.

16. For an excellent book-length treatment of hydration during exercise and exercise-associated hyponatremia, see Timothy Noakes, *Waterlogged* (Champaign, IL: Human Kinetics, 2012). See also Tamara Hew-Butler, "Arginine Vasopressin, Fluid Balance and Exercise: Is Exercise-Associated Hyponatraemia a Disorder of Arginine Vasopressin Secretion?," *Sports Medicine* 40 (2010): 459–479; Arthur J. Siegel, Joseph G. Verbalis, Stephen Clement, Jack H. Mendelson, Nancy K. Mello, Marvin Adner, Terry Shirey, Julie Glowacki, Elizabeth Lee-Lewandrowski, and Kent B. Lewandrowski, "Hyponatremia in Marathon Runners due to Inappropriate Arginine Vasopressin Secretion," *American Journal of Medicine* 120 (2007): 461. e11–461.e17.

17. Stuart Merson, Ronald Maughan, and Susan Shirreffs, "Rehydration with Drinks Differing in Sodium Concentration and Recovery from Moderate Exercise-Induced Hypohydration in Man," *European Journal of Applied Physiology* 103 (2008): 585–594; Kevin C. Miller, Gary W. Mack, and Kenneth L. Knight, "Gastric Emptying after Pickle-Juice Ingestion in Rested, Euhydrated Humans," *Journal of Athletic Training* 45 (2010): 601–608; Kevin C. Miller, Gary W. Mack, Kenneth L. Knight, J. Ty Hopkins, David O. Draper, Paul J. Fields, and Iain Hunter, "Reflex Inhibition of Electrically Induced Muscle Cramps in Hypohydrated Humans," *Medicine & Science in Sports & Exercise* 42 (2010): 953–961; Ernest Hooper, "Eagles' Juice Puts Dallas in a Pickle," *St. Petersburg Times,* September 5, 2000, http://www

.sptimes.com/News/090500/Sports/Eagles__juice_puts_Da.shtml (accessed May 2, 2012).

18. Jeffrey G. Wiese, Michael G. Shlipak, and Warren S. Browner, "The Alcohol Hangover," *Annals of Internal Medicine* 132 (2000): 897–902; Ruth M. Hobson and Ronald J. Maughan, "Hydration Status and the Diuretic Action of a Small Dose of Alcohol," *Alcohol and Alcoholism* 45 (2010): 366–373.

11. More Gain, Less Pain

1. Benjamin Franklin, "The Way to Wealth," in *A Benjamin Franklin Reader,* ed. Walter Isaacson (New York: Simon & Schuster, 2003), 176; Friedrich Nietzsche, *Twilight of the Idols,* trans. Duncan Large (New York: Oxford University Press, 1998), 5.

2. John J. McCarthy, Jyothi Mula, Mitsunori Miyazaki, Rod Erfani, Kelcye Garrison, Amreen B. Farooqui, Ratchakrit Srikuea, Benjamin A. Lawson, Barry Grimes, Charles Keller, Gary Van Zant, Kenneth S. Campbell, Karyn A. Esser, Esther E. Dupont-Versteegden, and Charlotte A. Peterson, "Effective Fiber Hypertrophy in Satellite Cell-Depleted Skeletal Muscle," *Development* 138 (2011): 3657–3666; T. Wessel, A. Haan, W. J. Laarse, and R. T. Jaspers, "The Muscle Fiber Type–Fiber Size Paradox: Hypertrophy or Oxidative Metabolism?," *European Journal of Applied Physiology* 110 (2010): 665–694; Michael J. Rennie, Henning Wackerhage, Espen E. Spangenburg, and Frank W. Booth, "Control of the Size of the Human Muscle Mass," *Annual Review of Physiology* 66 (2004): 799–828.

3. T. Garma, C. Kobayashi, F. Haddad, G. R. Adams, P. W. Bodell, and K. M. Baldwin, "Similar Acute Molecular Responses to Equivalent Volumes of Isometric, Lengthening, or Shortening Mode Resistance Exercise," *Journal of Applied Physiology* 102 (2007): 135–143.

4. Helen D. Kollias and John C. McDermott, "Transforming Growth Factor-β and Myostatin Signaling in Skeletal Muscle," *Journal of Applied Physiology* 104 (2008): 579–587; Alexandra C. McPherron and Se-Jin Lee, "Double Muscling in Cattle due to Mutations in the Myostatin Gene," *Proceedings of the National Academy of Sciences* 94 (1997): 12457–12461; Alexandra C. McPherron, Ann M. Lawler, and Se-Jin Lee, "Regulation of Skeletal Muscle Mass in Mice by a New TGF-β Superfamily Member," *Nature* 387 (1997): 83–90; Stephen Welle, Andrew Cardillo, Michelle Zanche, and Rabi Tawil, "Skeletal Muscle Gene Expression after Myostatin Knockout in Mature Mice," *Physiological Genomics* 38 (2009): 342–350; Helge Amthor, Raymond Macharia, Roberto Navarrete, Markus Schuelke, Susan C. Brown, Anthony Otto, Thomas Voit, Francesco Muntoni, Gerta Vrbóva,

Terence Partridge, Peter Zammit, Lutz Bunger, and Ketan Patel, "Lack of Myostatin Results in Excessive Muscle Growth but Impaired Force Generation," *Proceedings of the National Academy of Sciences* 104 (2007): 1835–1840.

5. Markus Schuelke, Kathryn R. Wagner, Leslie E. Stolz, Christoph Hübner, Thomas Riebel, Wolfgang Kömen, Thomas Braun, James F. Tobin, and Se-Jin Lee, "Myostatin Mutation Associated with Gross Muscle Hypertrophy in a Child," *New England Journal of Medicine* 350 (2004): 2682–2688; Dana S. Mosher, Pascale Quignon, Carlos D. Bustamante, Nathan B. Sutter, Cathryn S. Mellersh, Heidi G. Parker, and Elaine A. Ostrander, "A Mutation in the Myostatin Gene Increases Muscle Mass and Enhances Racing Performance in Heterozygote Dogs," *PLOS Genetics* 3 (2007): e79.

6. M. N. Fedoruk and J. L. Rupert, "Myostatin Inhibition: A Potential Performance Enhancement Strategy?," *Scandinavian Journal of Medicine & Science in Sports* 18 (2008): 123–131; Amelie Nadeau and George Karpati, "Are Big Muscles Necessarily Good Muscles?," *Annals of Neurology* 63 (2008): 543–545; "World Anti-Doping Agency–Prohibited List," http://www.wada-ama.org/ (accessed March 8, 2012).

7. Antonio Musaro, Karl McCullagh, Angelika Paul, Leslie Houghton, Gabriella Dobrowolny, Mario Molinaro, Elisabeth R. Barton, H. L. Sweeney, and Nadia Rosenthal, "Localized IGF-1 Transgene Expression Sustains Hypertrophy and Regeneration in Senescent Skeletal Muscle," *Nature Genetics* 27 (2001): 195–200.

8. The number of genes in the human genome is still not precisely known. Estimates have progressively decreased from about 100,000 genes to the current estimates of less than 25,000. "Human Genome Project Information—How Many Genes Are in the Human Genome?," http://www.ornl.gov/sci/techresources/Human_Genome/faq/genenumber.shtml (accessed March 8, 2012).

9. C. E. Stewart and J. M. Pell, "Point: IGF Is the Major Physiological Regulator of Muscle Mass," *Journal of Applied Physiology* 108 (2010): 1820–1821; Martin Flueck and Geoffrey Goldspink, "Counterpoint: IGF Is Not the Major Physiological Regulator of Muscle Mass," *Journal of Applied Physiology* 108 (2010): 1821–1823; Geoffrey Goldspink, "Mechanical Signals, IGF-I Gene Splicing, and Muscle Adaptation," *Physiology* 20 (2005): 232–238; Ronald W. Matheny, Bradley C. Nindl, and Martin L. Adamo, "Minireview: Mechano-growth Factor: A Putative Product of IGF-I Gene Expression Involved in Tissue Repair and Regeneration," *Endocrinology* 151 (2010): 865–875.

10. B. Hoier, N. Nordsborg, S. Andersen, L. Jensen, L. Nybo, J. Bangsbo, and Y. Hellsten, "Pro- and Anti-angiogenic Factors in Human Skeletal Muscle in Response to Acute Exercise and Training," *Journal of Physiology* 590 (2012): 595–606; Stuart Egginton, "Invited Review: Activity-Induced Angiogenesis," *Pflügers Archiv European Journal of Physiology* 457 (2009): 963–977.

11. There are several VEGF genes; this paragraph discusses VEGF-A, the one most involved in exercise-induced angiogenesis. Dawid G. Nowak, Jeanette Woolard, Elianna Mohamed Amin, Olga Konopatskaya, Moin A. Saleem, Amanda J. Churchill, Michael R. Ladomery, Steven J. Harper, and David O. Bates, "Expression of Pro- and Anti-angiogenic Isoforms of VEGF Is Differentially Regulated by Splicing and Growth Factors," *Journal of Cell Science* 121 (2008): 3487–3495; Steven J. Harper and David O. Bates, "VEGF-A Splicing: The Key to Anti-angiogenic Therapeutics?," *Nature Review Cancer* 8 (2008): 880–887.

12. PGC1 activates the transcription factor PPARgamma, hence the acronym PGC1 (PPARgamma coactivator). PPAR stands for peroxisome proliferator-activated receptor. Gamma identifies PPARgamma as one of three different PPARs. Peroxisomes are small cellular structures that break down fat molecules. PPARs, as their name suggests, were originally found to stimulate the synthesis of peroxisomes by serving as transcription factors. Subsequent work has identified many additional functions for PPARs.

13. PGC1 also has roles in other tissues, including liver and brown fat. In fat cells, upregulation of PGC1 promotes browning of white fat (see Chapter 8). Although PGC1 was named for its interactions with PPARgamma, PPARdelta is probably more important in skeletal muscle. Yong-Xu Wang, Chun-Li Zhang, Ruth T. Yu, Helen K. Cho, Michael C. Nelson, Corinne R. Bayuga-Ocampo, Jungyeob Ham, Heonjoong Kang, and Ronald M. Evans, "Regulation of Muscle Fiber Type and Running Endurance by PPARdelta," *PLOS Biology* 2 (2004): e294; Brian N. Finck and Daniel P. Kelly, "PGC-1 Coactivators: Inducible Regulators of Energy Metabolism in Health and Disease," *Journal of Clinical Investigation* 116 (2006): 615–622.

14. Jessica Maria Norrbom, Eva-Karin Sällstedt, H. Fischer, Carl Johan Sundberg, Helene Rundqvist, and Thomas Gustafsson, "Alternative Splice Variant PGC-1a-b Is Strongly Induced by Exercise in Human Skeletal Muscle," *American Journal of Physiology—Endocrinology and Metabolism* 301 (2011): E1092–E1098.

15. The quote is from Amby Burfoot, *The Runner's World Complete Book of Running: Everything You Need to Know to Run for Fun, Fitness, and Competition* (Emmaus, PA: Rodale Press, 1997), 153. The study on DNA methylation and diabetes is Romain Barrè, Megan E. Osler, Jie Yan, Anna Rune, Tomas Fritz, Kenneth Caidahl, Anna Krook, and Juleen R. Zierath, "Non-CpG Methylation of the PGC-1a Promoter through DNMT3B Controls Mitochondrial Density," *Cell Metabolism* 10 (2009): 189–198. For a highly readable introduction to the role of DNA methylation and similar mechanisms in health and disease see Nessa Carey, *The Epigenetics Revolution: How Modern Biology Is Rewriting Our Understanding of Genetics, Disease, and Inheritance* (London: Icon Books, 2011).

16. Romain Barrès, Jie Yan, Brendan Egan, Jonas Thue Treebak, Morten Rasmussen, Tomas Fritz, Kenneth Caidahl, Anna Krook, Donal J. O'Gorman, and Juleen R. Zierath, "Acute Exercise Remodels Promoter Methylation in Human Skeletal Muscle," *Cell Metabolism* 15 (2012): 405–411; Tina Rönn, Petr Volkov, Cajsa Davegårdh, Tasnim Dayeh, Elin Hall, Anders H. Olsson, Emma Nilsson, Åsa Tornberg, Marloes Dekker Nitert, Karl-Fredrik Eriksson, Helena A. Jones, Leif Groop, Charlotte Ling, "A Six Months Exercise Intervention Influences the Genome-wide DNA Methylation Pattern in Human Adipose Tissue," *PLOS Genetics* 9 (2013): e1003572.

17. Dongmei Liu, Maureen Sartor, Gustavo Nader, Laurie Gutmann, Mary Treutelaar, Emidio Pistilli, Heidi IglayReger, Charles Burant, Eric Hoffman, and Paul Gordon, "Skeletal Muscle Gene Expression in Response to Resistance Exercise: Sex Specific Regulation," *BMC Genomics* 11 (2010): 659.

18. Scientists distinguish small interfering RNA (siRNA) from microRNA (miRNA). Both are small regulatory RNAs, and their mechanisms are related, so I group them together under the term "microRNA" for the purposes of this discussion.

19. The protein structure denoted by the Research Collaboratory for Structural Bioinformatics Protein Data Bank's (RCSB's) Protein Data Base Identification (PDB ID): 2FFL represents a dicer from *Giardia intestinalis,* a protozoan parasite of the gut. Ian J. MacRae, Kaihong Zhou, Fei Li, Adrian Repic, Angela N. Brooks, W. Zacheus Cande, Paul D. Adams, and Jennifer A. Doudna, "Structural Basis for Double-Stranded RNA Processing by Dicer," *Science* 311 (2006): 195–198. PDB ID: 1U04 contains the structure of slicer (also called Argonaute) from a bacterium. Ar Ji-Joon Song, Stephanie K. Smith, Gregory J. Hannon, and Leemor Joshua-Tor, "Crystal Structure of Argonaute and Its Implications for RISC Slicer Ac-

tivity," *Science* 305 (2004): 1434–1437. See instructions in the Appendix for accessing and viewing these structures. To further explore the structure of slicer and dicer, see David Goodsell, "Molecule of the Month: Small Interfering RNA (siRNA)," http://dx.doi.org/10.2210/rcsb_pdb/mom_2008_2 (accessed August 7, 2012).

20. Monty Montano, "MicroRNAs: miRRORS of Health and Disease," *Translational Research* 157 (2011): 157–162; Robin C. Friedman, Kyle Kai-How Farh, Christopher B. Burge, and David P. Bartel, "Most Mammalian mRNAs Are Conserved Targets of MicroRNAs," *Genome Research* 19 (2009): 92–105.

21. Studies have examined responses in both humans and rodents: Søren Nielsen, Camilla Scheele, Christina Yfanti, Thorbjörn Åkerström, Anders R. Nielsen, Bente K. Pedersen, and Matthew Laye, "Muscle Specific MicroRNAs Are Regulated by Endurance Exercise in Human Skeletal Muscle," *Journal of Physiology* 588 (2010): 4029–4037; Adeel Safdar, Arkan Abadi, Mahmood Akhtar, Bart P. Hettinga, and Mark A. Tarnopolsky, "miRNA in the Regulation of Skeletal Muscle Adaptation to Acute Endurance Exercise in C57Bl/6J Male Mice," *PLOS One* 4 (2009): e5610; Isabelle Güller and Aaron P. Russell, "MicroRNAs in Skeletal Muscle: Their Role and Regulation in Development, Disease and Function," *Journal of Physiology* 588 (2010): 4075–4087.

22. Eva van Rooij, Daniel Quiat, Brett A. Johnson, Lillian B. Sutherland, Xiaoxia Qi, James A. Richardson, Robert J. Kelm, and Eric N. Olson, "A Family of MicroRNAs Encoded by Myosin Genes Governs Myosin Expression and Muscle Performance," *Developmental Cell* 17 (2009): 662–673.

23. Peter K. Davidsen, Iain J. Gallagher, Joseph W. Hartman, Mark A. Tarnopolsky, Flemming Dela, Jorn W. Helge, James A. Timmons, and Stuart M. Phillips, "High Responders to Resistance Exercise Training Demonstrate Differential Regulation of Skeletal Muscle MicroRNA Expression," *Journal of Applied Physiology* 110 (2011): 309–317.

24. Maxie Kohler, Wilhelm Schänzer, and Mario Thevis, "RNA Interference for Performance Enhancement and Detection in Doping Control," *Drug Testing and Analysis* 3 (2011): 661–667.

25. Nicholas Luden, Erik Hayes, Andrew Galpin, Kiril Minchev, Bozena Jemiolo, Ulrika Raue, Todd A. Trappe, Matthew P. Harber, Ted Bowers, and Scott Trappe, "Myocellular Basis for Tapering in Competitive Distance Runners," *Journal of Applied Physiology* 108 (2010): 1501–1509.

12. Chasing the Holy Grail

1. James S. Skinner, Artur Jaskólski, Anna Jaskólska, Joanne Krasnoff, Jacques Gagnon, Arthur S. Leon, D. C. Rao, Jack H. Wilmore, and Claude Bouchard, "Age, Sex, Race, Initial Fitness, and Response to Training: The HERITAGE Family Study," *Journal of Applied Physiology* 90 (2001): 1770–1776.

2. Claude Bouchard, E. Warwick Daw, Treva Rice, Louis PéRusse, Jacques Gagnon, Michael A. Province, Arthur S. Leon, D. C. Rao, James S. Skinner, and Jack H. Wilmore, "Familial Resemblance for VO2max in the Sedentary State: The HERITAGE Family Study," *Medicine & Science in Sports & Exercise* 30 (1998): 252–258.

3. David Sloan Wilson makes a similar argument about the influence of genes on criminal behavior in *Evolution for Everyone: How Darwin's Theory Can Change the Way We Think about Our Lives* (New York: Delacorte Press, 2007), 92–99.

4. David Epstein engagingly explains the HERITAGE study in *The Sports Gene: Inside the Science of Extraordinary Athletic Performance* (New York: Penguin Group, 2013); Skinner et al., "Age, Sex, Race," 1770–1776.

5. For an overview of polymorphisms in the human genome, see "U.S. Department of Energy Office of Science: SNP Fact Sheet," http://www.ornl.gov/sci/techresources/Human_Genome/faq/snps.shtml (accessed July 31, 2012); Feng Zhang, Wenli Gu, Matthew E. Hurles, and James R. Lupski, "Copy Number Variation in Human Health, Disease, and Evolution," *Annual Review of Genomics and Human Genetics* 10 (2009): 451–481. Key research articles include David A. Wheeler, Maithreyan Srinivasan, Michael Egholm, Yufeng Shen, Lei Chen, Amy McGuire, Wen He, Yi-Ju Chen, Vinod Makhijani, G. Thomas Roth, Xavier Gomes, Karrie Tartaro, Faheem Niazi, Cynthia L. Turcotte, Gerard P. Irzyk, James R. Lupski, Craig Chinault, Xing-zhi Song, Yue Liu, Ye Yuan, Lynne Nazareth, Xiang Qin, Donna M. Muzny, Marcel Margulies, George M. Weinstock, Richard A. Gibbs, and Jonathan M. Rothberg, "The Complete Genome of an Individual by Massively Parallel DNA Sequencing," *Nature* 452 (2008): 872–876; David G. Wang, Jian-Bing Fan, Chia-Jen Siao, Anthony Berno, Peter Young, Ron Sapolsky, Ghassan Ghandour, Nancy Perkins, Ellen Winchester, Jessica Spencer, Leonid Kruglyak, Lincoln Stein, Linda Hsie, Thodoros Topaloglou, Earl Hubbell, Elizabeth Robinson, Michael Mittmann, Macdonald S. Morris, Naiping Shen, Dan Kilburn, John Rioux, Chad Nusbaum, Steve Rozen, Thomas J. Hudson, Robert Lipshutz, Mark

Chee, and Eric S. Lander, "Large-Scale Identification, Mapping, and Genotyping of Single-Nucleotide Polymorphisms in the Human Genome," *Science* 280 (1998): 1077–1082.

6. M. Vainzof, C. S. Costa, S. K. Marie, E. S. Moreira, U. Reed, M. R. Passos-Bueno, A. H. Beggs, and M. Zatz, "Deficiency of Alpha-actinin-3 (ACTN3) Occurs in Different Forms of Muscular Dystrophy," *Neuropediatrics* 28 (1997): 223–228.

7. The protein structure denoted by the Research Collaboratory for Structural Bioinformatics Protein Data Bank's (RCSB's) Protein Data Base Identification (PDB ID): 108A represents human ACE with an inhibitor bound to it. Ramanathan Natesh, Sylva L. U. Schwager, Edward D. Sturrock, and K. Ravi Acharya, "Crystal Structure of the Human Angiotensin-Converting Enzyme–Lisinopril Complex," *Nature* 421 (2003): 551–554. See instructions in the Appendix for accessing and viewing this structure. Note the globular shape of this enzyme.

8. B. Rigat, C. Hubert, F. Alhenc-Gelas, F. Cambien, P. Corvol, and F. Soubrier, "An Insertion/Deletion Polymorphism in the Angiotensin I-Converting Enzyme Gene Accounting for Half the Variance of Serum Enzyme Levels," *Journal of Clinical Investigation* 86 (1990): 1343–1346; Molly S. Bray, James M. Hagberg, Louis PéRusse, Tuomo Rankinen, Stephen M. Roth, Bernd Wolfarth, and Claude Bouchard, "The Human Gene Map for Performance and Health-Related Fitness Phenotypes: The 2006–2007 Update," *Medicine & Science in Sports & Exercise* 41 (2009): 35–73.

9. H. E. Montgomery, R. Marshall, H. Hemingway, S. Myerson, P. Clarkson, C. Dollery, M. Hayward, D. E. Holliman, M. Jubb, M. World, E. L. Thomas, A. E. Brynes, N. Saeed, M. Barnard, J. D. Bell, K. Prasad, M. Rayson, P. J. Talmud, and S. E. Humphries, "Human Gene for Physical Performance," *Nature* 393 (1998): 221–222; Malcolm Collins, Stavroulla L. Xenophontos, Marios A. Cariolou, Gaonyadiwe G. Mokone, Dale E. Hudson, Lakis Anastasiades, and Timothy D. Noakes, "The ACE Gene and Endurance Performance during the South African Ironman Triathlons," *Medicine & Science in Sports & Exercise* 36 (2004): 1314–1320.

10. A. G. Williams, M. P. Rayson, M. Jubb, M. World, D. R. Woods, M. Hayward, J. Martin, S. E. Humphries, and H. E. Montgomery, "Physiology: The ACE Gene and Muscle Performance," *Nature* 403 (2000): 614–614.

11. Yannis P. Pitsiladis, Guan Wang, and Bernd Wolfarth, "Genomics of Aerobic Capacity and Endurance Performance: Clinical Applications," in *Exercise Genomics,* ed. L. S. Pescatello and S. M. Roth (New York: Humana Press, 2011), 179–230.

12. Heather Gordish-Dressman and Joseph M. Devaney, "Statistical and Methodological Considerations in Exercise Genomics," in *Exercise Genomics,* ed. L. S. Pescatello and S. M. Roth (New York: Humana Press, 2011), 23–44; Corneliu Henegar, Henri Hooton and Karine Clément, "Search for Genes Influencing Complex Traits," in *Genetic and Molecular Aspects of Sport Performance,* ed. C. Bouchard and E. P. Hoffman (Hoboken, NJ: Wiley-Blackwell, 2011), 70–78.

13. Treva K. Rice, Mark A. Sarzynski, Yun Ju Sung, George Argyropoulos, Adrian M. Stütz, Margarita Teran-Garcia, D. C. Rao, Claude Bouchard, and Tuomo Rankinen, "Fine Mapping of a QTL on Chromosome 13 for Submaximal Exercise Capacity Training Response: The HERITAGE Family Study," *European Journal of Applied Physiology* 112 (2012): 2969–2978.

14. Genome-wide association studies are compiled at "National Human Genome Research Institute—A Catalog of Published Genome-wide Association Studies," http://www.genome.gov/gwastudies/ (accessed July 30, 2012).

15. Claude Bouchard, Mark A. Sarzynski, Treva K. Rice, William E. Kraus, Timothy S. Church, Yun Ju Sung, D. C. Rao, and Tuomo Rankinen, "Genomic Predictors of the Maximal O_2 Uptake Response to Standardized Exercise Training Programs," *Journal of Applied Physiology* 110 (2011): 1160–1170.

16. The current cost is below $10,000 and falling according to "National Human Genome Research Institute—Genome Sequencing Costs," http://www.genome.gov/sequencingcosts/ (accessed July 30, 2012).

17. Philippe Houle-Leroy, Helga Guderley, John G. Swallow, and Theodore Garland, "Artificial Selection for High Activity Favors Mighty Mini-muscles in House Mice," *American Journal of Physiology—Regulatory, Integrative and Comparative Physiology* 284 (2003): R433–R443.

18. For the record, all of the players I mention contributed to the Bulls' success. You can reach your own conclusions about their relative contributions by examining their statistics at "NBA and ABA Player Directory," http://www.basketball-reference.com/players (accessed July 30, 2012).

Appendix

1. "Research Collaboratory for Structural Bioinformatics Protein Data Bank," http://www.rcsb.org/pdb/ (accessed August 7, 2012).

2. These files are assigned a four-character code that designates the particular protein, followed by the .pdb suffix (e.g., 1CAG.pdb).

3. John Moreland, Apostol Gramada, Oleksandr Buzko, Qing Zhang, and Philip Bourne, "The Molecular Biology Toolkit (MBT): A Modular Platform for Developing Molecular Visualization Applications," *BMC Bioinformatics* 6 (2005): 21; Dong Xu and Yang Zhang, "Generating Triangulated Macromolecular Surfaces by Euclidean Distance Transform," *PLOS One* 4 (2009): e8140; and "Jmol: An Open-Source Java Viewer for chemical Structures in 3D," http://jmol.sourceforge.net/ (accessed August 7, 2012).

4. Search for "RCSB PDB" in the app store for your device.

5. J. Bella, M. Eaton, B. Brodsky, and H. M. Berman, "Crystal and Molecular Structure of a Collagen-like Peptide at 1.9 A Resolution," *Science* 266 (1994): 75–81; Sushmita D. Lahiri, Pan-Fen Wang, Patricia C. Babbitt, Michael J. McLeish, George L. Kenyon, and Karen N. Allen, "The 2.1 Å Structure of *Torpedo californica* Creatine Kinase Complexed with the ADP-Mg2+ NO3- Creatine Transition-State Analogue Complex," *Biochemistry* 41 (2002): 13861–13867.

6. "RCSB PDB Help System," http://www.pdb.org/pdb/staticHelp.do ?p=help/index.html (accessed August 7, 2012). An excellent tutorial for RCSB Protein Workshop is linked to this source.

7. David S. Goodsell, "Molecule of the Month Archive," http://www .rcsb.org/pdb/101/motm_archive.do (accessed August 7, 2012).

8. "Foldit: Solve Puzzles for Science," http://fold.it/portal/ (accessed August 8, 2012). Foldit users have contributed to real scientific progress. For example, see Firas Khatib, Frank DiMaio, Foldit Contenders Group, Foldit Void Crushers Group, Seth Cooper, Maciej Kazmierczyk, Miroslaw Gilski, Szymon Krzywda, Helena Zábranská, Iva Pichová, James Thompson, Zoran Popović, Mariusz Jaskolski, and David Baker, "Crystal Structure of a Monomeric Retroviral Protease Solved by Protein Folding Game Players," *Nature Structural and Molecular Biology* 18 (2011): 1175–1177.

Glossary

Acetate A two-carbon molecule produced by metabolic pathways that break down both sugars and fats. In cells, acetate is often attached to coenzyme A to form acetyl CoA.

Acetylcholine Neurotransmitter released at the neuromuscular junction and many other synapses.

Actinin-3 gene (ACTN3) Gene that codes for the protein actinin-3. Variations in ACTN3 are correlated with performance in events requiring power.

Action potential Electrical signal generated in neurons and used to encode and carry information along their axons.

Adenosine diphosphate (ADP) Product formed when ATP breaks down.

Adenosine triphosphate (ATP) Key molecule of energy metabolism. The energy released when ATP breaks down into ADP and phosphate powers many cellular processes.

Adrenalin Hormone released during stressful conditions, including exercise, triggering increased heart rate, mobilization of sugar and lipid reserves, and many other effects. Also called epinephrine.

Aerobic metabolism Metabolic pathways that require oxygen to extract energy from nutrients.

Allele Alternate form of a gene located at a specific place along a chromosome. Alleles differ from each other in the sequence of nucleotides.

Allosteric regulation Change in the activity of an enzyme or other protein due to the binding of molecules to regulatory sites on the protein.

Amino acid Single carbon molecules that are linked together to form proteins. Amino acids have variable side chains that differ in shape, charge, reactivity, and other properties, resulting in twenty amino acids with distinct biochemical properties.

Anaerobic metabolism Metabolic pathways that do not require oxygen to extract energy from nutrients.

Angiotensin-converting enzyme (ACE) Enzyme that converts the hormone angiotensin to an active form. Variation in the gene encoding ACE is associated with endurance.

Axon Long projection from the cell body of a neuron. Axons carry information around the body in the form of action potentials.

Brain-derived neurotrophic factor (BDNF) A molecule produced in the brain that encourages growth of neurons.

Bulk flow Movement of fluid or air driven by pressure gradients.

Channel Proteins that permit specific molecules to pass through cell membranes in a regulated manner.

Chemical energy Energy contained in atoms, molecules, and chemical bonds. Chemical energy can be absorbed, transferred, or released during chemical reactions.

Chromosome Structure formed by a long molecule of DNA bound to proteins. Genes are located along the linear sequence of nucleotides in chromosomal DNA.

Citric acid cycle Cycle of chemical reactions in mitochondria that extracts energy from acetate, generating high energy electron carriers that eventually are used to synthesize ATP.

Coding region Portion of a gene that contains instructions for making a protein.

Concentric contraction Muscle contraction during which the muscle produces force while shortening.

Cotransporter Membrane protein that simultaneously carries two or more molecules across a cell membrane.

Covalent bond Strong chemical bonds in which two atoms share electrons.

Covalent regulation Activation or inactivation of enzymes or other proteins by attachment of molecules with covalent bonds.

Crossbridge Connection between the head group of a myosin molecule and one subunit of an actin filament. Movements of crossbridges produce force during muscle contractions.

Cytoskeleton Web of proteins in cells that provides structure, mediates movements, and regulates other processes.

Diffusion Movement of molecules down concentration or electrochemical gradients due to the molecule's random movements.

Duty ratio Portion of time that muscle myosin molecules spend attached to actin. Duty ratio varies across muscle types.

Eccentric contraction Muscle contraction during which the muscle produces force while lengthening.

Electrochemical gradient Combined influences of electrical charge and concentration on the tendency of a molecule to move across a cell membrane.

Electron carrier Molecule that temporarily binds an electron and delivers it to another molecule. In mitochondria, electron carriers link nutrient breakdown to ATP synthesis.

Enzymes Proteins and other biomolecules that accelerate specific chemical reactions. Some enzymes also link chemical reactions to each other.

Epithelial layer A single layer of cells that separates the inside and outside of the body. Epithelial layers mediate movement of molecules into and out of the body.

Erythropoietin (EPO) Hormone released by the kidney that stimulates synthesis of new red cells.

Extend Movement that increases the angle between bones, usually resulting in the straightening of a limb.

Extensible Capable of being extended without breaking.

Fatty acids Molecules composed mainly of long chains of carbon with attached hydrogens. Fatty acids are the main component of biological lipids.

Fermentation The breakdown of carbon-containing compounds in the absence of oxygen. Lactate fermentation is the conversion of pyruvate to lactate, both compounds with three carbons.

Flex Movement that decreases the angle between bones, usually resulting in the bending of a limb.

Gastrocnemius Muscle composed of two large units that together with the smaller soleus muscle forms the calf.

Gene Basic unit of heredity, composed of a string of nucleotides along a stretch of DNA. Genes include instructions to synthesize functional products, including proteins and RNAs, along with regulatory elements.

Genomewide survey Studies that evaluate the correlation between some measure of health or performance and variation in genes across the entire genome.

Glycogen Molecule composed of many glucoses linked together. Glycogen stores glucose in readily available form mainly in the liver and muscle.

Glycolysis Metabolic pathway that breaks down six-carbon glucose molecules to two three-carbon pyruvate molecules, with a net synthesis of two ATP molecules.

Heterozygous The condition in which the two copies of a particular gene carried by an individual are different alleles.

Homeostasis Maintenance of stable internal temperature, acidity, osmolarity, and other variables in the face of a changing external environment.

Homozygous Condition in which the two copies of a particular gene carried by an individual are the same allele.

Hydrogen bond Bonds between partial positive and partial negative charges on nearby atoms. Hydrogen bonds are weaker than covalent bonds.

Hydrophobic Unable to easily dissolve in water. Hydrophobic molecules tend to clump together in water.

Hypothalamus Region of the brain with a vital role in hormone secretion and homeostasis of temperature, osmolarity, and other variables.

Hypoxia-inducible factor (HIF) A transcription factor that is active when oxygen levels fall.

Insulin Small protein hormone secreted by the pancreas that stimulates uptake of glucose from the blood into cells of the liver, muscle, fat, and other tissues and promotes synthesis of storage molecules including glycogen and fatty acids.

Insulin-like growth factor (IGF) Small protein with structural similarity to insulin that promotes muscle growth.

Interneuron Neuron in the central nervous system that connects one neuron to another.

Ion Charged molecule, formed when a neutral atom gains or loses an electron.

Isometric contraction Muscle contraction during which the muscle produces force while remaining the same length.

Kinase Enzyme that catalyzes the attachment of a phosphate group to another molecule.

Kinetic energy Energy possessed by an object due to its motion.

Lactate threshold Exercise intensity at which lactate begins to accumulate at high levels in the blood.

Long-term potentiation (LTP) Long-lasting enhancement of the strength of a synapse.

Maximal aerobic capacity (VO_{2max}) Measure of the peak amount of oxygen that a person or other organism can consume during exercise.

Membrane Thin barrier composed of phospholipids that defines the outer surface of cells and encloses some structures within cells. Molecules that are large or charged do not easily cross membranes without assistance from membrane proteins.

Messenger ribonucleic acid (mRNA) mRNA molecules carry sequence information from DNA to the ribosome, where protein synthesis occurs.

Micrograph A magnified digital image or photograph taken through a microscope.

MicroRNA Very short ribonucleic acid (RNA) chains that have roles in regulating the synthesis of proteins from mRNA.

Mitochondria Intracellular structures enclosed in two membranes that are the sites of ATP production during aerobic metabolism.

Motor neuron Neurons that carry information from the central nervous system to muscles and other tissues.

Motor unit Motor neuron and all of the muscle fibers it regulates.

Muscle fiber A muscle cell.

Myosin Proteins with a globular head group, a short neck, and a long tail that convert the chemical energy of ATP into movements through their interactions with actin and other molecules.

Myostatin Protein secreted mainly by muscle cells that inhibits muscle growth.

Negative feedback pathway Regulatory system that acts to maintain a variable at a constant level.

Neuromuscular junction Synapse that connects motor neurons to muscle fibers.

Neuron Cell of the nervous system that transmits information through chemical and electrical signals.

Neurotransmitter Molecule that carries signals from one neuron to another across a synapse.

Noradrenalin Neurotransmitter released during stressful conditions, including exercise, triggering increased heart rate, mobilization of sugar and lipid reserves, and many other effects. Also called norepinephrine.

Nucleotides Small molecules that contain a sugar, phosphate group, and a base. Nucleotides link together to form DNA and RNA. ATP is a modified version of the nucleotide adenosine.

Osmolarity Measure of the concentration of dissolved solute in a fluid.

Osmotic gradient Difference between the osmolarity of two solutions. Water tends to flow across membranes and other barriers from low to high osmolarity.

Oxidative phosphorylation Last portion of aerobic metabolism in which ATP is synthesized from ADP and oxygen is consumed.

Paralogs Different genes in the same species that have similar but not identical nucleotide sequences and that typically code for proteins with somewhat divergent functions.

Phosphatase Enzyme that removes an attached phosphate group from a molecule.

Phosphate Ion that contains a phosphorus and three oxygen atoms and carries three negative charges.

Polymorphisms Variants of a gene's DNA sequence that occur with high frequency in populations.

Positive feedback Process in which small changes or perturbations become amplified.

Potassium Chemical element that occurs mainly as positively charged potassium ions in biological fluids.

Potential energy Energy that results from the position of an object or the arrangement of particles within the object. Forms of potential energy include gravitational, pressure, chemical, and elastic potential energy.

Power Amount of energy used or transferred per unit time. When a force is moving an object, power is the product of force and velocity.

P_{oxygen} Partial pressure of oxygen. In air, the amount of pressure contributed by oxygen to the total pressure atmospheric pressure. In solutions, both the concentration and solubility of oxygen influence its partial pressure. Oxygen diffuses along its partial pressure gradient.

PPAR gamma coactivator (PGC) Protein that cooperates with transcription factors to regulate the expression of specific genes. PGC1 levels rise during endurance exercise and contribute to the endurance-training response.

Pre-RNA Initial molecule produced when a portion of the DNA in a gene is copied to RNA.

Proteasome Structures inside cells that degrade damaged and unnecessary proteins.

Proteins Molecules with complex three-dimensional shapes that are composed of linked amino acids. Proteins regulate cellular processes, provide structural support, mediate movements of cells and their components, generate force, and catalyze chemical reactions.

Proton Positively charged ion of a hydrogen atom. High-proton concentrations correspond to high acidity and low pH. Sometimes abbreviated as H^+.

Pump Membrane protein that uses the energy from ATP breakdown to transport molecules across cell membranes, establishing concentration and electrical gradients.

Pyruvate Three-carbon product of glucose breakdown by glycolysis. Pyruvate can be converted to acetate and enter the citric acid cycle or be converted to lactate by fermentation.

Receptor Protein that receives a chemical, electrical, or other signal and transits the signal to regulatory pathways inside a cell.

Ribonucleic acid (RNA) Chains of linked nucleotides. RNAs carry information, catalyze reactions, and regulate protein synthesis and other processes.

Sarcomere Basic functional unit of muscles cells, composed of interdigitated actin-rich thin filaments and myosin-rich thick filaments.

Sarcoplasmic reticulum Membrane-bound structure inside muscle cells that sequesters calcium ions during muscle relaxation and releases them to trigger contractions.

Sensory neuron Neuron that detects a stimulus and carries it toward the central nervous system.

Single-nucleotide polymorphisms (SNPs) Polymorphisms resulting from mutations in a single nucleotide.

Skeletal muscle Muscle that attaches to and moves bones.

Sodium Chemical element that occurs mainly as positively charged sodium ions in biological fluids.

Solute Dissolved molecule.

Spliceosome Complex of both protein and RNA that removes regions that do not code for protein from pre-RNA molecules.

Splice variants Variants in mRNA encoded by the same gene. Splice variants result from differences in processing of pre-RNA molecules.

Synapses Structures that pass information from one neuron to another. In chemical synapses, one nerve cell releases a neurotransmitter that binds to receptors on the other neuron, triggering responses.

Tendon Band of connective tissue rich in the protein collagen that transmits force from muscles to bones.

Thick filament Strands that are rich in myosin protein and interdigitate with thin filaments to form sarcomeres.

Thin filament Strands that are rich in actin protein and interdigitate with thick filaments to form sarcomeres.

Transcription factor Proteins that bind to regulatory regions on genes and influence the synthesis of RNAs that encode specific products.

Transporter Membrane protein that carries molecules across cell membranes by binding them, changing its shape, and then releasing them on the other side of the membrane. Transporters do not directly use ATP.

Type 1 muscle fibers Slow-twitch muscle cells that are fatigue resistant but not capable of producing high forces.

Type 2A muscle fibers Muscle fibers with fatigue resistance and force production that are intermediate between Type 1 and Type 2X fibers.

Type 2X muscle fibers Fast-twitch muscle cells that fatigue easily but are capable of producing high forces. Also called Type 2B muscle fibers.

Uncoupling protein (UCP) Protein located on the inner membrane of mitochondria that allows protons to pass without synthesizing ATP.

Vascular endothelial growth factor (VEGF) Small protein that stimulates growth of new blood vessels and promotes proliferation of brain cells.

Vasopressin Hormone that causes water to be reabsorbed from the urine, resulting in urine with low volume and high salt concentration, and causes blood vessels to constrict, increasing blood pressure. Also called antidiuretic hormone (ADH).

Vesicle Small membrane-bound sphere that can fuse with the cell membrane and release its contents.

Voltage Potential energy gradient resulting from differences in charge across the membrane. Voltage can drive charge across a membrane in the same way that blood pressure can move fluid through a vessel.

Acknowledgments

After a race, I often reflect about the people who helped me before, during, and after the run: spectators, volunteers, organizers, other racers, training partners, family members, doctors, and physical therapists. Though superficially solitary, distance running is a collaborative activity. So too, I have found over the past two years, is writing a book.

Scientific research is also a cooperative enterprise. The insights described in this book represent the work of thousands of scientists. Though my emphasis has been on the science rather than the scientists, I hope that readers will share my appreciation for the commitment and brilliance of the scientists whose work I describe. My apologies to many whose work I did not have space to mention. I owe a special debt to those who mentored me during my training: Charles Holliday, Bliss Forbush III, Gary Mack, and the late Ethan Nadel. Each pushed me to excel while encouraging me to find my own path.

My deepest debt is to my wife, Kathy, for reasons too numerous to fully describe here. Here's the beginning of the list: Kathy read each chapter multiple times, pointed me toward many terrific sources, cleaned up some very messy grammar, and patiently listened to countless analogies, helping me refine the promising ones and reject the many clunkers. Family, friends, and colleagues also offered support, encouragement, and lots of constructive criticism. G. P. Gillen and Ron Dukes gave me feedback on drafts of each chapter. John Hofferberth commented on two chapters and taught me quite a bit of chemistry in the process. Andrew Kerkhoff, Stephanie Berg, Peter Gillen, and Irene Gillen gave feedback on portions of the book. The comments of two anonymous reviewers greatly improved this manuscript. Many of the stories I tell and the connections to athletics I describe first emerged during runs with friends. I am grateful for all the encouragement and advice I received, but of course, any errors are my responsibility.

Michael Fisher at Harvard University Press planted the idea for this project many years before I was ready to pursue it. I appreciate his enthusiasm and wise advice throughout the project. I am also grateful to the many others at HUP and Westchester Publishing Services who helped bring this book to completion, including Katherine Brick, Lauren Esdaile, and John Donohue.

I wrote this book from an office in the Kenyon Athletic Center while on sabbatical leave. When I craved inspiration, I needed only to peer outside my door to see athletes in action. I truly appreciate the generosity of Peter Smith and many others at the KAC who offered me not only an office, but also a supportive and lively atmosphere for writing. I launched this project as a participant in the *Kenyon Review* Writer's Workshop, or perhaps more accurately, the workshop launched me into this project. My thanks to instructors Dinty Moore and Jenny Patton, to the other participants in my workshop group, and to the many others who made the workshop an incredible opportunity to grow as a writer.

I am most fortunate to work with terrific colleagues and students at Kenyon College. I am deeply grateful to my students, whose experiences, knowledge, and feedback have taught me much about exercise physiology and how to teach it effectively. Kenyon colleagues inside and outside the biology department inspired this project and offered helpful advice along the way. Kenyon's support of a sabbatical leave made this project possible.

Index